£3.00

SUPERCONDUCTIVITY AND ITS APPLICATIONS
J.E.C.Williams

Applied physics series

SUPERCONDUCTIVITY AND ITS APPLICATIONS
J.E.C.Williams

Pion Limited, 207 Brondesbury Park, London NW2

© 1970 Pion Limited

SBN 85086 010 5

Set on IBM 72 Composers by Pion Limited, London.
Printed in Great Britain by J.W.Arrowsmith Limited, Bristol.

Preface

Since 1960 there has been, in the modern idiom, an information explosion in the field of superconductivity. It is the purpose of this present monograph to sift the debris of this explosion and to piece together the aspects of superconductivity upon which its most significant applications are based.

There are many excellent books dealing either with the intricacies of the microscopic theory of superconductivity or with its phenomenological treatment. However, for the newcomer to superconductivity, particularly for the applications engineer, there exists little by way of a simple introduction to the basic physical causes of superconductivity and its application to practical devices. This book seeks to remedy this lack. The theory underlying the practical applications of superconductivity is stressed rather than the theory of superconductivity itself.

In the development of the theory generalities are, wherever possible, avoided and specific examples used. For example, in the analysis of flux penetration, one-dimensional equations only are employed, from which very simple solutions are obtained. Mechanical models of physical phenomena are used wherever possible. Most of the fine detail of the microscopic and phenomenological theories has been omitted where it was felt that it would not have contributed to the understanding of the applications of superconductivity.

Superconductivity is one of the most spectacular and remarkable effects of the microcosmic physical universe. Because it touches on many aspects of solid-state physics, an understanding of superconductivity gives an insight into some of the complexities of the solid state. It is hoped that this book will help to unveil some of the abstractions of the subject and to reveal some of the burgeoning possibilities that arise from it.

J.E.C.Williams

For Ann

Contents

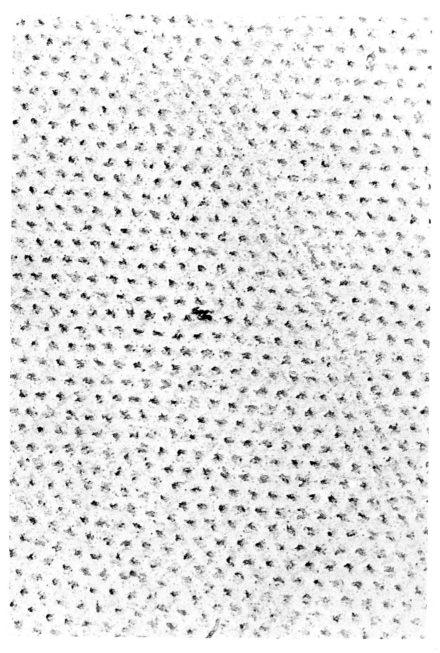

Plate I. The powder reproduction patterns of the Abrikosov flux array in a type II superconductor. Dislocations can be seen in the pattern similar to the dislocations in a crystal lattice. The material in this case is Pb–6·3 at.% In and the field strength is 56 A cm^{-1}. The magnification is ×8300.
Courtesy of Dr. U. Essman, Max Planck Institut, Stuttgart.

A survey of superconductivity

1.1 Introduction

Although progress in the understanding and application of superconductivity has been steady if slow since 1911, a rapid acceleration in the pace occurred in 1960. For it was in that year that the first of a number of 'high-field' superconductors was discovered.

The property of zero electrical resistance is that which comes immediately to mind in thinking of superconductors and it is this property alone which is exploited in the most widespread of all superconductive devices, the superconducting magnet.

The generation of an intense magnetic field (one that is greater than the saturation field of ferromagnets) is an operation unique in engineering practice: it is performed at zero efficiency. For, although no energy is expended by a static magnetic field, energy must be expended to create it. For instance, in a water-cooled copper magnet generating 120 kA cm^{-1} in a 5 cm bore, 5 MW of power are dissipated in the copper. The advent of superconductors capable of remaining superconductive in high magnetic fields enabled the generation of fields up to 120 kA cm^{-1} for only a few watts of power. The application of superconductors to magnet construction far outweighs in number and commercial significance all other applications of superconductivity at the present time.

Attempts to exploit 'high-field' superconductivity in low-frequency a.c. devices such as alternators and transformers have not so far met with great success, although small superconductive a.c. power devices have been built. The reason for this frustration is the resistivity of 'high-field' superconductors under a.c. conditions.

However, a class of superconductors, known generically as type I, is essentially without resistance at frequencies up to 100 MHz. This property has enabled very-high-frequency devices, such as resonant cavity accelerators (linear accelerators, 'linacs' for short) to be built with values of magnification up to 10^{10}. Such 'linacs' can operate continuously with only a fraction of the power requirements of conventional accelerators.

The current in a type I superconductor is carried in a very thin surface layer. Although the current density is very high, $\sim 10^8$ A cm^{-2}, the layer is shallow, $\sim 10^{-5}$ cm, so that surface currents of only 1000 A cm^{-1} can be carried. Furthermore, type I superconductors remain superconducting only in comparatively low fields, up to 800 A cm^{-1}. In contrast high-field superconductors, otherwise known as type II can, in special circumstances, carry bulk current densities as high as 10^6 A cm^{-2} at fields of up to 120 kA cm^{-1}. Thus type I superconductors may be used generally for low-field low-current-density high-frequency applications whilst type II superconductors can be used for high-field d.c. devices.

Because the current flow in a type I superconductor is confined to the surface, no dissipative losses occur when the current flow patterns change. Because the surface currents are intimately associated with the incident field it follows that a type I superconductor can be moved without friction through a magnetic field. This property of type I superconductors has been exploited in the 'frictionless' bearing and the superconductive gyroscope, in which a superconducting sphere is levitated and spun in a magnetic field.

Superconductivity occurs only in certain materials and at low temperatures. At a particular transition temperature, which depends on the material, on the ambient magnetic field, and on the transport current, if any, a change from superconductive to normal resistive behaviour occurs. The range of temperature over which the change occurs depends mainly on the purity of the material. In some pure metals, such as tin, the range is extremely narrow, perhaps no more than $10^{-3}\,^\circ$K. The effective temperature coefficient of resistance is therefore extremely high and it can be exploited in a bolometer. Radiation levels as low as 10^{-12} W can be detected by the superconductive tin bolometer.

The rapid variation of resistance with magnetic field strength at the transition field can be used as a mechanism of amplification, and thus a family of electronic devices can be developed. These include amplifiers and oscillators and bistable circuits.

The latter are an important class of superconductive device depending for their operating principle on the duality of states of a superconductor on either side of its magnetic transition point. The cryotron is essentially a switch formed from a pair of superconducting amplifiers. By thin-film evaporation techniques the cryotron can be made exceedingly small. It is being actively developed as a high-speed computer element.

The ability of a superconductor to store circulating current indefinitely can be used as the basis of a computer memory unit. Circulating currents can be stored in a porous evaporated film of tin or indium in a particular spatial array symbolising the binary digits 1 or 0. This device is called the continuous film memory and is being developed for possible superconducting computers.

A unique property of superconductivity is what is known as flux quantisation. The magnetic flux threading a closed superconducting current loop can only take on integral multiples of 2×10^{-15} Wb. This remarkable and important characteristic can be exploited directly and indirectly in a number of quantum-measuring devices. These include the absolute fluxmeter, absolute ammeter, and absolute voltmeter.

In the absolute fluxmeter and its associated device, the absolute ammeter, the number of flux quanta entering or leaving a superconducting loop are counted. The process of counting is achieved either by a weak superconducting link in the circuit or by a device called a Josephson junction. The latter can also measure voltage in terms of frequency, for under particular conditions a voltage may be developed across a Josephson junction which is irradiated by high-frequency radiation. The voltage is an absolute and linear function of the frequency.

This brief review of the devices which are being developed from superconductivity will serve to show what a rapidly expanding field of science it is.

Understanding the basic physics of any subject, but particularly of superconductivity, is largely a matter of 'learning the language'; the engineering application consists mainly of 'thinking of things to say'. It is hoped that this small book will lead to a basic vocabulary from which thought can be constructed.

1.2 Units

Because the intention of this book is to bridge the gap between basic physics and engineering application, it is appropriate to draw attention to the units that will be used.

The design of superconducting devices, as of all other products of electrical engineering, uses for its basic units volts, amperes, joules, and consequently all the S.I. units. These will be used throughout this book.

However, physicists in particular and many engineers frequently use c.g.s. units, especially in specifying field strengths and flux densities, and to both of these the unit of gauss is loosely applied. The relationship between c.g.s. and S.I. units of magnetic measurement are as follows:

flux $\qquad$ 1 Wb $\quad = 10^8$ lines $= 10^8$ G cm^2,
flux density $\quad$ 1 T $\qquad = 1$ Wb m$^{-2} = 10^4$ G,
field strength $\quad$ 1 A m$^{-1} = 0 \cdot 01259$ Oe.

In order to make the mental conversion of field strength between c.g.s. and S.I. units somewhat easier than it appears above, we shall usually quote field strengths in amperes per centimetre. This can be easily multiplied mentally by $1\frac{1}{4}$ to give the field strength in oersteds, for the c.g.s. user. The product may even be called 'gauss', if units are used particularly carelessly. However, whenever field strength is quoted in amperes per centimetre, it must be converted to amperes per metre before substitution into any equation.

Similarly, resistivity is quoted in ohm centimetres because the values are universally known. For instance, the resistivity of copper at room temperature is generally quoted as $1 \cdot 76 \times 10^{-6}$ Ω cm. But for use in expressions in this book it is converted to $1 \cdot 76 \times 10^{-8}$ Ω m.

The history of superconductivity

2.1 The discovery of soft superconductors[†]

Soon after Kammerlingh Onnes liquefied helium in 1908 a range of temperatures below $4\cdot2°$K became accessible for experiment. In 1911 Onnes investigated the temperature variation of the electrical resistance of mercury in this range. The rather surprising result was the total disappearance of resistance at $4\cdot15°$K, although Onnes anticipated that such an effect might occur. By using large measuring currents he showed that an upper bound to the resistivity below this temperature was 10^{-14} Ω cm. He concluded that a significant change occurred in the state of this metal below $4\cdot15°$K and dubbed it the superconducting state.

The upper bound to the resistivity in this state was quickly lowered to 10^{-20} Ω cm, this being determined by measuring the decay of persistent current in a superconducting ring over a period of many months. There is now sufficient evidence to justify the assumption of zero resistance at least in one type of superconductor. Indium, tin, and lead were soon added to the list of superconductors and it was erroneously thought for a time that superconductivity was a peculiarity of the soft metals.

Onnes's discovery immediately raised the hope of building a superconducting magnet, a hope that was short lived, however (Keesom, 1935; Rjabinin and Schubnikov, 1935). It was found that comparatively small magnetic fields quenched superconductivity. For instance, lead, whose resistivity in zero magnetic field disappears below $7\cdot2°$K, is superconducting at $4\cdot2°$K only for field strengths of less than about 480 A cm⁻¹.

Furthermore, it was found that the current that quenched superconductivity in a round wire varied as its diameter, not as its cross-sectional area, so that high current densities could only be obtained in very thin wires. Indeed Silsbee (1916) showed that the current which quenched superconductivity in a round wire was merely that current which generated the quenching field at the surface.

The nearest prewar approach to the construction of a superconducting magnet of useful proportions occurred in 1930 with the rather surprising discovery that the eutectic alloy of lead and bismuth remained superconducting in fields of up to about 12 kA cm⁻¹ at $4\cdot2°$K. However, the current density in the alloy was low and, although it now appears that the construction of a 1 T magnet might have been feasible, attempts to construct one were abandoned (Keesom, 1935; Rjabinin and Schubnikov, 1935).

In 1929 it was suggested that the change in state between the normal and the superconducting states might be treated thermodynamically just

† Onnes (1911).

as the change in state at melting or boiling, or as changes in crystal structure or any other reversible process may be treated (Gorter, 1933).

Indeed the thermodynamic treatment of the observable phenomena proved highly successful. Most of the developments in the theory of superconductivity between 1930 and 1950 consisted of refinements to the thermodynamics of the superconducting–normal transition. The success of the thermodynamic treatment indicated that the transition was a reversible one, proof of which was obtained in 1933 by Meissner and Ochsenfeld.

But it was not until 1957 that superconductivity was explained in terms of microscopic solid-state phenomena; then the theory of Bardeen, Cooper, and Shrieffer, 1957 (to be referred to as BCS) was published. What this theory did, which no previous theory had done, was to reconcile two fundamental and apparently conflicting characteristics of superconductors, the critical temperature and the critical field strength.

2.2 The basic properties of superconductors†

The superconducting state of a metal exists only in a particular range of temperature and field strength. The condition for the superconducting state to exist in the metal is that some combination of temperature and field strength should be less than a critical value. (In the practical application of superconductors in magnets, transport current may be added to the list of conditional parameters which define the superconducting region. However, in considering simple elemental superconductors it will be seen that current density is indistinguishable from field strength and is not therefore an independent conditional parameter.)

The critical values of field strength and temperature are found experimentally to be closely related by the equation

$$H_c = H_0 \left[1 - \left(\frac{T}{T_c} \right)^2 \right], \qquad (2.2.1)$$

where H_c is the critical field strength at the temperature T, H_0 is the maximum critical field strength occurring at absolute zero, and T_c is the critical temperature, the highest temperature for superconductivity. Thus Equation (2.2.1) defines a curve which divides the normal region of the field–temperature diagram of the metal from the superconducting region.

The critical field curves for a number of pure metals are shown in Figure 2.2.1. It is from these curves that the surprising paradox of superconductivity arises. We shall consider it with reference to lead. At the absolute zero of temperature the critical field of lead is 640 A cm⁻¹. This means that, in the absence of lattice vibration, which has ceased at absolute zero, an increase in the energy content of the metal equivalent to a field of 640 A cm⁻¹ quenches superconductivity. This energy density is 240 J m⁻³ or 10⁻⁷ eV atom⁻¹.

† Schoenberg (1952).

By contrast, in zero magnetic field strength, the critical temperature of lead is $7 \cdot 175°K$, which corresponds to a lattice energy of about 8×10^{-4} eV atom^{-1}.

Thus we see that the magnetic energy required to raise the electrons of lead from the superconducting to the normal state is only one-tenthousandth of that required to normalise the superconducting state by thermal energy. This is the paradox that any successful theory of superconductivity must explain.

To understand the origin of this phenomenon we must first briefly review the quantum electronic model of a metal.

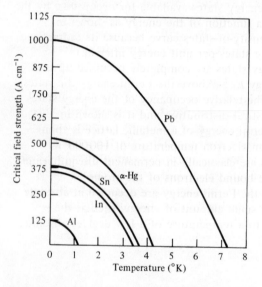

Figure 2.2.1. Curves of critical field strength as a function of absolute temperature for a number of pure, metallic superconductors.

The superconducting energy gap

3.1 The microscopic causes of superconductivity†

A metal consists of a lattice of regularly located atoms. Although macroscopically the lattice is electrically neutral, not all the electrons are bound to the lattice atoms. These free electrons can move through the lattice influenced by the potential of the ionised lattice atoms. The distribution of energies (velocities) of these free electrons is such that each electron has a discrete and unique energy. It is thus said to occupy one energy state. The number of energy states available for occupation by the free electrons of the lattice is a function of the energy as shown in Figure 3.i.1. (It is called a density-of-states curve because its ordinates represent the number of energy states per unit energy interval.)

At absolute zero these energy states are completely occupied up to a level known as the Fermi energy E_F. Above the Fermi energy the states are completely unoccupied. The relative occupancy of the energy states is called the Fermi–Dirac statistical distribution and it is shown in Figure 3.1.2. Typically the Fermi energy of a metallic lattice is about 10 eV, which corresponds to an electron temperature of $100\,000°$K. Clearly then the free electrons are classically in permanent disequilibrium with the lattice, just as are the bound electrons of the atoms.

Because all the states up to the Fermi energy are occupied at absolute zero, the application of a very small amount of energy, such as the thermal energy corresponding to a temperature of a few degrees absolute,

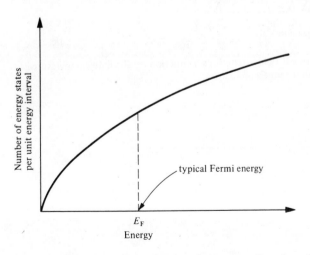

Figure 3.1.1. Curves of the density of states as a function of electron energy for a normal metal.

† Little (1965).

can only affect the electrons very close to the Fermi energy. The Fermi–
Dirac distributions at temperatures above absolute zero are shown in
Figure 3.1.2. Some electrons now occupy states above the Fermi energy
and some of the states just below it are unoccupied. The Fermi–Dirac
distribution at absolute zero can also be changed by the application of a
voltage impulse as shown in Figure 3.1.3. The preponderance of electrons
with positive velocity indicates a current flow. At absolute zero and in a
perfect lattice this current flow will be resistanceless.

These processes in which the Fermi–Dirac distribution is changed can
be compared with the excitation of an electron from the bound state in
an atom to the free, conduction state. Unlike this latter case, however,
an electron in a metallic lattice can be excited from a 'bound' state just
below the Fermi energy to a 'free' state just above it by the supply of an
arbitrarily small energy. It is this that accounts for a zero of electronic

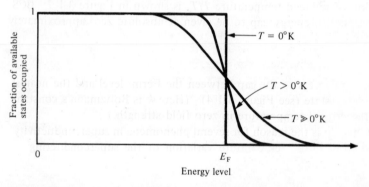

Figure 3.1.2. The Fermi–Dirac distribution function at absolute zero and at finite
temperatures. The amplitude of this function at a particular energy level indicates
the fraction of available energy states at that level that will be occupied by
electrons.

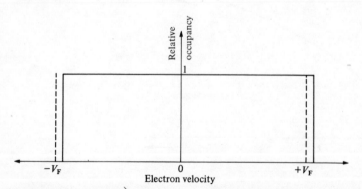

Figure 3.1.3. The Fermi–Dirac distribution modified asymmetrically by a voltage
impulse. The impulse has given all the electrons a positive component of velocity
thus creating a current.

specific heat of a metal at absolute zero. At temperatures above absolute zero the states below E_F become depleted and those above E_F become filled as shown in Figure 3.1.4. Thus a greater energy must be supplied to an electron to raise it to an available energy state. This manifests itself as a linear increase in electronic specific heat with temperature (Lynton, 1964).

The energy spectrum of the free electrons of a material in the superconducting state differs from that of the same material in the normal state by the appearance of a gap in the density of states centred about the Fermi energy. The width of this gap is typically 10^{-4} to 10^{-3} eV. It can be measured in various ways, the simplest being by electron tunnelling in which the essential measuring instrument is a voltmeter (Giaver, 1961). This experiment is described later in Section 3.2.

The gap depends on both temperature and field strength. Its variation as a function of reduced temperature T/T_c is shown in Figure 3.1.7. BCS theory predicts this energy gap to be given at absolute zero approximately by the expression

$$2\epsilon_0 = 3 \cdot 5 k T_c$$

where ϵ_0 always refers to the gap between the Fermi level and the nearest available energy state (see Figure 3.1.4). (Here k is Boltzmann's constant and T_c is the critical temperature in zero field strength.)

The energy gap is the reason for several phenomena in superconductivity such as an exponential temperature variation of the superconducting

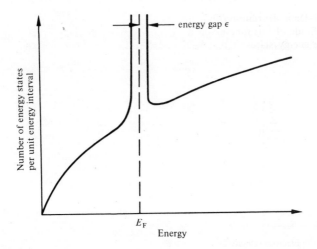

Figure 3.1.4. The density-of-states curve of a metal in the superconducting state. Energy levels which are normally located in the region of the gap are piled up on either side of it in an abnormally dense arrangement.

electronic specific heat, the occurrence of a threshold potential for electron tunnelling, and most important, the persistent current.

How this energy gap arises we shall now see.

Let us turn to a model of a lattice, being traversed by a pair of electrons as shown in Figure 3.1.5. As electron A moves past the lattice ions it alters their positions by Coulomb attraction creating a locally increased positive charge density.

This increase in positive charge density is not purely local, for the movement of the lattice ions radiates away from the centre of the disturbance with the velocity of sound. This radiating disturbance takes the form of a spectrum of phonons, each phonon being a quantised packet of acoustic energy. As each phonon propagates through the lattice it will encounter other electrons with which it will interact. Thus through the intermediacy of the phonons the electrons can experience mutual attractive forces at a distance. The average maximum distance at which the phonon coupled attraction can occur is called the coherence length and is given the symbol ξ.

In particular it is helpful to consider a pair of electrons moving through the lattice along a common circular (or at least closed) orbit. If the electrons are pictured as being mutually diametrically opposite and having equal but opposite momentum, a special situation arises. For now, through the action of the phonons radiated by each towards the other, the two electrons experience a continuing mutual attraction. Now whether or not electrons

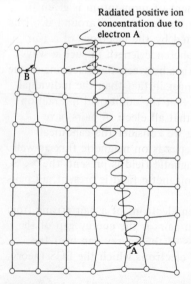

Radiated positive ion
concentration due to
electron A

Figure 3.1.5. Conceptual representation of a lattice being traversed by a pair of electrons, each emitting phonons and being influenced by them. Only one of an enormous number of phonons is shown propagating through the lattice.

actually circulate in that way is immaterial. Indeed, as will be shown later, the concept of individual electrons is of no use in describing several properties of superconductors. However, insofar as a mechanical model is helpful in grasping abstract ideas, the picture of a mutually circulating pair of electrons is offered as a guide.

It can be understood now that movement of the electron pair relative to the lattice will change the effect of the phonon interaction. For, as the phonons travel only at the velocity of sound, the movement of the electrons through the lattice at velocities approaching that will reduce the attractive forces to the point at which the pairs are no longer held together. This happens at the surface of a superconductor under the influence of the critical magnetic field.

So far we have considered only one pair of electrons and only those phonons radiated by each electron which interact with the other pairs of electrons in their vicinity.

The direct mutual interaction of the two electrons of a pair (called a Cooper pair) is an attractive interaction which lowers the energy of the pair relative to the mean energy (Fermi energy) of unpaired electrons. The energy is lowered because a finite amount of work must be done to separate them.

The magnitude of this energy is indicated by the critical field strength of the superconductor, for this field just accelerates the electron pairs to a velocity at which the phonon interaction ceases. The magnetic field affects all electron pairs simultaneously and, although a common momentum is imparted to the electron pairs, no component of momentum is given to any particular pair to differentiate it from the sea of pairs around it. This fact is important. Now the phonons emitted by the electrons of a pair interact with other electrons within the coherence length in such a way that an additional attractive interaction exists between the electrons of the pair. This additional attraction is far stronger in fact than the individual phonon attraction. However, a condition that the additional interaction should be attractive and not disruptive is that all electron pairs have a common momentum, and for this purpose we consider the canonical momentum of the electron pairs, which depends on magnetic flux as well as on velocity (see Section 6.1). If no magnetic field is present, the momentum of all pairs will be zero: if a magnetic field is present, all pairs will have the same canonical momentum.

The additional strong attraction between electrons of a pair arising from the collective interaction of many electron pairs is the energy gap of the superconductor. It is the very large excess of the energy gap over the binding energy of a single isolated pair of electrons which the BCS theory so successfully predicts.

For most superconductors the energy gap is a factor of 10^3 to 10^4 larger than the correlation energy of an isolated electron pair.

Let us now consider the perturbations which may disrupt a superconducting electron pair. A magnetic field accelerates the pairs at the surface of a superconductor (see Section 5.1) and gives to *all* the pairs, including those lying deep within the superconductor and therefore not actually accelerated, a common canonical momentum. Because all the pairs have the same momentum, the interaction between them, and hence the energy gap, is unaltered. If the electron pairs at the surface are accelerated too close to the velocity of sound, however, pairing is disrupted and the energy gap falls abruptly to zero. That at least is what should in theory happen. In fact, for reasons not well understood, the energy gap is steadily reduced by a rising magnetic field, as shown in Figure 3.1.6.

Increasing temperature reduces the energy gap. It does so for the following reason. The temperature of a lattice is synonymous with a spectrum of phonons propagating continuously throughout the lattice and causing random movement of the lattice atoms. As the temperature rises the amplitude and frequency of movement of the lattice atoms increases, and these movements interfere with the propagation of the phonons between correlated electron pairs. This clearly results in a decrease in the attractive interaction between pairs with a consequent decrease in the energy gap. The variation of the energy gap as a function of temperature is shown in Figure 3.1.7. At the absolute zero of temperature the energy gap is given by

$$2\epsilon_0 = 3 \cdot 5 k T_c \,, \tag{3.1.1}$$

where ϵ_0 is the energy gap, that is the difference between the Fermi energy and the nearest available energy state, T_c is the critical temperature in zero field, and k is Boltzmann's constant.

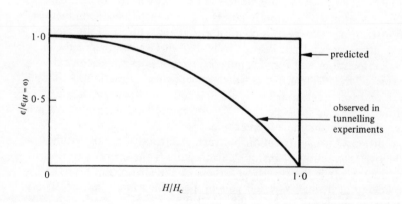

Figure 3.1.6. The variation of the energy gap as a function of magnetic field strength, in theory and as observed.

Near to the critical temperature the variation of the energy gap can be expressed approximately as a function of temperature by

$$\epsilon = 3 \cdot 2kT_c \left(1 - \frac{T}{T_c}\right)^{1/2}. \tag{3.1.2}$$

Using the microscopic theory, Equation (2.2.1) can be more accurately written as

$$H_c = H_0 \left[1 - 1 \cdot 07 \left(\frac{T}{T_c}\right)^2\right] \tag{3.1.3}$$

From this brief qualitative description of the microscopic super-conducting state we can now define two types of perturbation by which superconductivity can be destroyed. Those perturbations which affect the electron pairs collectively require only low energy densities to destroy superconductivity, for they need only exceed the correlation energy. Examples of such perturbations are magnetic fields and voltage impulses.

Those perturbations which affect electron pairs individually must be of greater energy density to destroy superconductivity, for they must exceed the energy gap. Such perturbations are the random lattice movements caused by temperature and radiation quanta.

Radiation of a frequency of about 10^{11} Hz is strongly absorbed by superconductors. This arises from the equivalence of the energy of a photon of this frequency to the energy gap. At higher frequencies the photon energy exceeds the energy gap so that the superconductor behaves more like a normal conductor. 10^{11} Hz is below the near-infrared and visible spectrum frequencies, which explains why metals in the super-conducting state look like normal metals and have the same low-temperature emissivity.

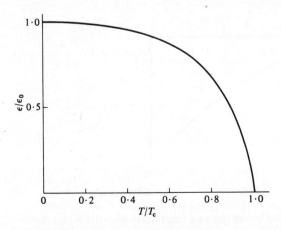

Figure 3.1.7. The variation of the energy gap as a function of temperature. This theoretical curve is confirmed experimentally.

3.2 Superconducting electron tunnelling†

An interesting consequence of the energy gap in superconductors is the
phenomenon called tunnelling.   In superconductors tunnelling manifests
itself as a flow of superconducting electrons between a superconductor and
another metal, either normal or superconducting, through an interposed
thin insulating film.

The structure of a typical tunnel junction is shown in Figure 3.2.1.
It consists of a lead film evaporated onto a glass substrate, a superimposed
layer of lead oxide of about 100 Å thickness and finally an evaporated
film of tin. Connections are made to the lead and tin films so that the
potential between them can be measured as a function of the current
passing between them. At $4\cdot2°K$ the lead is superconducting but the tin
is normal: under these conditions the voltage–current characteristic is as
shown in Figure 3.2.2. If the temperature of the sandwich is lowered to
$3°K$, at which both lead and tin are superconducting, the voltage–current
characteristic is as shown in Figure 3.2.3. The shape of these curves is a
consequence of the energy gap in the superconductors.

The spatial distribution of electrons in a metal can be represented by a
wavefunction in which the probability of the existence of an electron (or
electron pair) at a particular position at a particular time is represented
by the amplitude of the wavefunction at that point.   The only characteristic
of the wavefunction for superconducting electron pairs that need be
considered at the present time is the finite spatial rate of change at the
boundary of a superconductor. Because of this, there is a small, but finite

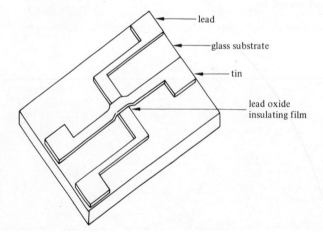

Figure 3.2.1. The structure of a tunnel junction. The junction consists of two
metals, one or both being a superconductor, separated by a thin oxide or other
insulating layer. The lead and tin layers are consecutively evaporated through masks.

† Giaver (1961).

probability of the existence of electrons beyond the physical boundaries
of the superconductor. Thus, when a superconductor and another metal
(either normal or superconducting) are placed in close proximity, some
electrons of each metal can exist in the other.

Consider Figure 3.2.4. It shows the energy spectra (that is, the product
of the density-of-states curve and the Fermi–Dirac distribution) for a

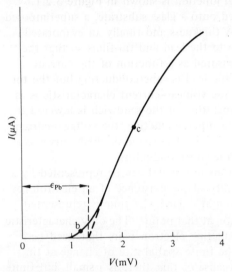

Figure 3.2.2. The voltage as a function of current for a tunnel junction consisting of
superconducting lead and normal tin.

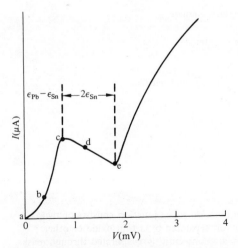

Figure 3.2.3. The voltage as a function of current for a tunnel junction consisting of
superconducting lead and superconducting tin.

superconductor and a normal metal at a few degrees absolute. The curves are displayed in pairs in such a way as to depict the electronic conditions prevailing in, for instance, superconducting lead and normal tin on each side of an insulating layer. In each material it is seen that some of the states available above the Fermi energy are occupied, and some of the states below it are unoccupied.

The unoccupied states are available to be filled by electrons no matter where they originate, provided only that the energy of the incoming electron exactly matches that of the state it is to occupy.

Suppose the Fermi energies of the superconducting lead and normal tin are level. The situation is then as shown in Figure 3.2.4a. Electrons in the region A above the energy gap in the lead can tunnel into the almost totally unoccupied band of energies in the tin (region B). The rate at which the electrons tunnel is a function of the number available in

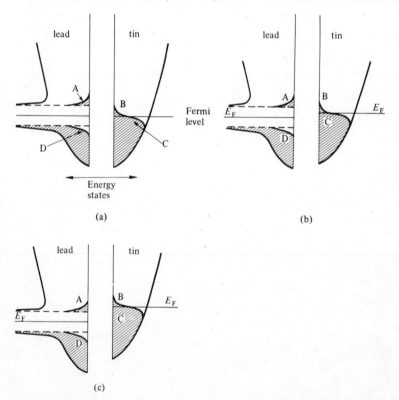

Figure 3.2.4. The energy spectrum curves for two metals, one superconducting and one normal. The shaded regions are each the product of the density-of-states curve and the Fermi–Dirac distribution. They represent therefore the number of occupied states at a given energy level. The two curves are set against a common energy axis to show the relative numbers of electrons in the superconductor and normal metal at each energy level.

region A of the lead and the number of unoccupied states available in region B of the tin. At the same time electrons below the Fermi energy in the tin (region C) can tunnel into the partly empty energy states below the energy gap in the lead (region D). Although far fewer empty states are available in region B than in D, many more electrons are available to tunnel from region C than from A. Thus the rates of tunnelling from lead to tin and tin to lead are equal. This equilibrium condition obtains only when the Fermi energies of the two metals are level, although of course the Fermi energies of the metals need not be equal. They may be made level by the electrical connection of the two metals or by asymmetric tunnelling until equilibrium is reached. How the latter occurs can be seen by considering Figure 3.2.4b. In this case the Fermi energies are not level in the two metals either because a potential difference is maintained between the metals by a battery, or because the lead or the tin was initially electrically charged. Suppose in fact the tin was initially charged negatively so that its Fermi energy was higher than that of the lead. Now electrons can still tunnel from the occupied states in region A of the lead to the unoccupied states of region B of the tin. However, this tunnel current is lower than before because fewer unoccupied states in the tin lie level with the occupied states of region A. On the other hand, the tunnel current from C to D is increased because more occupied states in C lie level with unoccupied states in D. Thus there is a small net current flow from tin to lead. This rapidly restores the charge balance of the two metals and equilibrium returns. If the potential difference is maintained as by a battery, the net tunnel current will persist and the relevant point on the current–voltage characteristic will be point b of Figure 3.2.2. The larger part of the occupied states in the tin above the Fermi energy lie opposite the energy gap in the lead; therefore these electrons cannot tunnel from the tin in this region.

Suppose now, however, the potential difference between the tin and the lead is increased so that the relative levels of the energy spectra are as shown in Figure 3.2.4c. Now the electrons occupying states above the Fermi energy in the tin can tunnel into the largely unoccupied region A in the lead. This results in a rapid rise in the tunnel current which is now represented by point c of Figure 3.2.2. As the potential difference is still further increased, many more occupied states in region C lie opposite the empty states in A, and the current rises ever more rapidly.

Thus it is seen that the main tunnel current is due to the coincidence of largely occupied and largely unoccupied states in the two metals. On the other hand, the initial small current flowing at small potential differences is due to the slight asymmetry of states caused by the parabolic variation of the density of states coupled with the gap in the states available for occupation by electrons close to the Fermi energy in the tin. The rapid rise in tunnel current occurs when the potential difference is half the energy gap.

The presence of occupied states above the Fermi energy and of unoccupied states below it is due to the assumed finite temperature. At absolute zero there are no 'tails' to the Fermi distribution. The voltage–current characteristic then has a sharper rise when the potential difference is one-half the energy gap.

Before considering the tunnelling processes between two superconductors it is important to put approximate values to the voltages involved. Lead, with a critical temperature of $7 \cdot 175°$K has an energy gap at absolute zero given by

$$2\epsilon_0 = 3 \cdot 5k \times 7 \cdot 175 = 2 \cdot 16 \times 10^{-3} \text{ eV}.$$

At a temperature of, say, $3°$K

$$2\epsilon = 1 \cdot 5 \times 10^{-3} \text{ eV}.$$

At the same temperature the typical spread of energies in the occupied levels above the gap and the unoccupied levels below the gap is

$$\epsilon' \approx kT = 0 \cdot 26 \times 10^{-3} \text{ eV}.$$

The Fermi energy of lead is about 10 eV. Thus the 'smearing' of the Fermi–Dirac distribution at the finite temperature of $3°$K is much less than the width of the gap, and both are infinitesimal in comparison with the Fermi energy.

With these relative magnitudes in mind let us suppose that the temperature of the lead–lead-oxide–tin sandwich is reduced to $3°$K so that the tin is also superconducting. There being initially no potential difference between the lead and the tin, the energy spectra of the two metals will be as shown in Figure 3.2.5a. If the potential difference now rises slightly, a net tunnelling current flows from the tin to the lead, because, as before, fewer electrons in the filled levels at A move into the unoccupied levels at B than tunnel from C to D. This situation is represented by the point b in Figure 3.2.3.

As the potential difference is increased and the energy levels of the tin are raised relative to those of the lead, the net tunnel current increases until the energy levels at the top of each gap coincide. At this juncture the tunnel current has reached a temporary maximum, shown as point c in Figure 3.2.3. For now, if the potential of the tin is further raised with respect to the lead, fewer occupied states in C lie opposite the partly occupied states in D. At the same time the current flow from B to A increases but less rapidly than the flow from C to D decreases. Hence the net tunnel current decreases with increasing potential difference. This stage is represented typically by point d of Figure 3.2.3. The fall of tunnel current continues until the bottom of the energy gap of the tin is level with the top of the energy gap of the lead, a situation represented by Figure 3.2.5c. Then, with further increase in potential difference the heavily occupied states in the tin (region C) become level with the

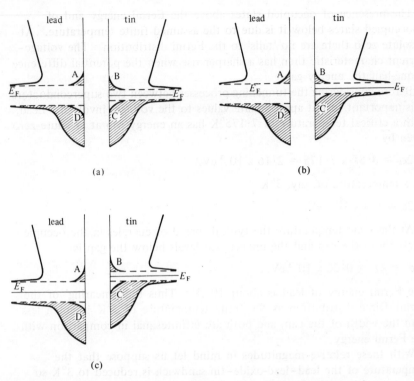

Figure 3.2.5. The energy spectrum curves for two superconducting metals.

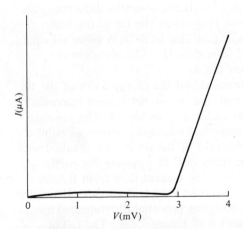

Figure 3.2.6. The tunnelling current as a function of voltage difference for two superconductors at temperatures very near to absolute zero. The negative resistance characteristic has nearly disappeared because there are only very small 'tails' to the energy spectra.

unoccupied states in the lead (region A). The tunnel current at this juncture is represented by point e of Figure 3.2.3. As in the previous case the tunnel current then increases rapidly. The width of the region of negative dI/dV is $2\epsilon_{tin}$, that is, about $1 \cdot 1$ mV.

At absolute zero or temperatures very close to it, the curve of tunnel current as a function of voltage is as shown in Figure 3.2.6. In that case no current flows until the potential difference is $\epsilon_{lead} + \epsilon_{tin}$.

Although the 'negative resistance' characteristic of tunnelling between superconductors could be exploited in a variety of devices, its main use has been in the measurement of the energy gaps of superconductors.

In essence it allows the measurement to be made directly, with a voltmeter.

The occurrence of superconductivity

4.1 The occurrence of superconductivity†

Superconductivity occurs in many elements, compounds, and alloys. Among the elements it occurs in two main groups in the periodic table: (1) the transition elements; (2) the groups IIB, IIIA, IVA. Superconductivity does not occur (as far as is presently known) in ferromagnetic elements, in insulators, in the rare earths, in groups IA, IIA, in semiconducting elements, nor in group IIB. The known superconducting elements are listed in Table 4.1.1, together with their principal critical parameters. The entries against niobium and vanadium are explained in Section 7.3. Most of the common elements found to be non-superconductive have been examined in fields of less than 1 A cm^{-1} to temperatures of 0·05°K. Although only twenty-six elements are known to be superconductors, the number of superconducting compounds is well over three hundred. Some of the binary compounds contain two superconducting elements (for example Nb_3Sn), some contain only one (for example V_3Si), and some contain none (for example $SrBi_3$). In addition there are a multitude of superconducting alloys such as Nb–Ti, Nb–Zr, Mo–Re, and Pb–Bi. The properties of the compound Nb_3Sn and the alloy Nb–Ti are shown in Tables 9.5.1 and 9.5.2.

No theory yet predicts definitely the occurrence of superconductivity in the elements, although criteria for superconductivity in terms of the Fermi velocity and the sound velocity of metals have been proposed which indicate that only some metals should be superconductors (Kulik, 1965).

We can understand qualitatively, however, why the good normal conductors such as copper, silver, and sodium are not superconducting (Langenberg et al., 1966). We have seen that a prerequisite for super-conductivity is a strong acoustic coupling between the electrons and the lattice. Good normal conductors by contrast are such mainly because they have weak interaction between the electrons and the lattice, and it is this which accounts mainly for their lack of superconductivity. Indeed we shall see later that a criterion for pronounced superconductivity in one class of superconductors is a very low normal electrical conductivity.

Empirical rules exist with which the likelihood of the occurrence of superconductivity can be judged and with which the approximate critical temperatures of new materials can be extrapolated from those of known superconductors. These criteria, known as the Matthias rules, can be summarised briefly as follows.

Superconductivity is favoured by a large atomic volume (large lattice parameter) and by a particular average number of valence electrons per

† Matthias (1963).

atom. In transition elements, alloys, and compounds, superconductivity occurs only for between two and eight valence electrons per atom. The relative critical temperature of elemental superconductors is shown as a function of the number of valence electrons per atom in Figure 4.1.1.

Table 4.1.1. The superconducting elements.

Element	T_c (°K)	H_0 (A cm^{-1})	H_{c2} (0) (A cm^{-1})	κ [a]	Reference
Aluminium	1·191	79			1958, *Phys. Rev.*, **111**, 132
Cadmium	0·52	26			1952, *Phys. Rev.*, **88**, 1172
Gallium	1·087	43			1961, *Phys. Rev.*, **121**, 1688
Hafnium	0·165	–			1963, *Rev. Mod. Phys.*, **35**, 1
Indium	3·408	228			1960, *Phys. Rev.*, **120**, 88
Iridium	0·14	16			1962, *Phys. Rev. Letters*, **8**, 408
Lead	7·175	640			1961, *Phys. Rev.*, **122**, 1888
Lanthanum α	4·9	800			1958, *Phys. Rev.*, **109**, 70, 243
Lanthanum β	5·9	800			
Mercury α	4·153	328			1961, *Phys. Rev.*, **123**, 1115
Mercury β	3·949	270			
Molybdenum	0·93	78			1963, *Phys. Rev.*, **129**, 136
Niobium	9·2	1600	2400	1·5	1962, *Phys. Rev. Letters*, **9**, 371
Osmium	0·71	52			1957, *Phys. Rev.*, **106**, 659
Rhenium	1·70	160			1957, *Phys. Rev.*, **106**, 659
Ruthenium	0·47	37			1957, *Phys. Rev.*, **106**, 859
Tantalum	4·48	660			1960, *Phys. Rev.*, **120**, 88
Technetium	8·22	–			1962, *Phys. Rev. Letters*, **9**, 254
Tellurium (at pressures exceeding 56000 atm.)					1964, *Phys. Rev. Letters*, **13**, 202
Thallium	2·39	136			1954, *Phys. Rev.*, **95**, 333
Thorium	1·37	129			1956, *Phil. Mag.*, **3**, 591
Tin	3·72	246			1960, *Phys. Rev.*, **120**, 88
Titanium	0·39	16			1953, *Phys. Rev.*, **89**, 654
Tungsten	0·01	0·16			1964, *Phys. Rev. Letters*, **12**, 688
Uranium	0·68	110			1957, *Phys. Rev.*, **107**, 1517
Vanadium	5·13	1030	5600	3·5	1952, *Phys. Rev.*, **85**, 85
Zinc	0·85	42			1958, *Phys. Rev.*, **112**, 1083
Zirconium	0·55	246			1960, *Phys. Rev.*, **120**, 88

[a] For the significance of κ, see Section 7.3.

Transition elements, alloys, and compounds having the same atomic volumes will have critical temperatures roughly as indicated by Figure 4.1.1.

Having thus briefly stated the basic cause of superconductivity we can proceed to consider the thermodynamics of the phase transition.

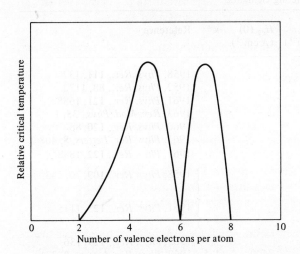

Figure 4.1.1. The relative critical temperature of superconductors as a function of the number of valence electrons. These approximate curves are known as the Matthias rule.

The thermodynamics of type I superconductors

5.1 The thermodynamics of the phase change†

Materials, usually but not always metals, which are superconductive can exist in either the superconducting or the normal state. A number of the physical properties of the materials are different in these two states, notably, of course, the electrical resistivity. However, the specific heat is also different as is the thermal conductivity and, to a very small extent, the mechanical properties.

The superconducting state can only exist at very low temperatures, typically in the range below $20°K$, although only one material is presently known to be superconducting at $20°K$. More are superconducting below $10°K$, still more below $1°K$. These criteria of temperature apply in zero magnetic field. If a magnetic field surrounds a material, the temperature below which it is superconducting is decreased.

Suppose a superconductor is immersed in a bath of liquid helium boiling at $4·2°K$ at atmospheric pressure and a magnetic field is applied. Because the superconductor has no resistance, the rising magnetic field generates surface currents which exclude the magnetic flux from the centre of the superconductor and which cannot decay through resistive dissipation. At some critical field, denoted by H_c, the shielding current density reaches a limiting value and the magnetic field penetrates the bulk of the superconductor which returns to the normal state. This is a magnetic transition of the first order, so called because there is a discontinuity of the first derivative of free energy with respect to the field at the transition.

For a superconductor in the shape of a long rod parallel to the field and therefore of negligible demagnetising coefficient, the transition from the completely superconducting to the completely normal state takes place over a very narrow range of field strengths.

The critical field for the transition depends on the temperature, being greatest at absolute zero at which it is denoted H_0. At the critical temperature T_c of the superconductor the critical field is zero. The variation of critical field as a function of temperature for a number of pure metals is shown in Figure 2.2.1.

The current required to quench superconductivity in a round wire is called the Silsbee current. It is that current which just produces the critical field at the surface. Thus,

$$I_c = 2\pi H_c r ,\qquad(5.1.1)$$

where r is the radius of the wire in metres and I_c is the critical current in amperes. H_c is then measured in amperes per metre.

† Schoenberg (1952).

It may be appreciated that the penetration of the magnetic field into the wire is an unstable situation. For, as the current retreats into the remaining superconducting portion of the wire, the circumscribing field strength rises. In fact the magnetic flux penetrates in a spatially periodic pattern. This spatially periodic penetration is a characteristic of superconducting samples of finite demagnetising coefficient; it is called the intermediate state and is dealt with briefly in Section 11.2 (Schoenberg, 1952).

One of the earliest successes in the theoretical treatment of superconductivity was the application of thermodynamics to the phase transition.

Thermodynamics can be applied to the analysis of this phase change because it is reversible, like the melting of a solid or the excitation of an atom. This reversibility was suspected early in the development of superconductivity but was not proved experimentally until after the successful development of the thermodynamic theory.

This theory concerns the relationship between the magnetic and thermal properties of a superconductor at the phase transition. It is developed as follows.

As the strength of the magnetic field surrounding the superconductor rises, shielding currents build up on the surface and the superconductor thereby achieves a certain negative magnetisation, i.e. the field strength inside is less than that outside. For soft superconducting metals like lead or mercury more than 1 μm in size, this magnetisation is given by $M = -H$; and this magnetisation increases the free energy of the superconductor. The shielding currents arise in a manner described in Section 5.2.

The free energy of a material arises from its thermodynamic state. Thus, a material's temperature, entropy, pressure, magnetisation, and even its electric polarisation determine its free energy. But, in considering the phase transition of a superconductor to the normal state at constant temperature, only the change in magnetisation is significant.

A body in which the magnetic flux density is lower than in the free space around it is magnetised negatively, and the difference in flux densities gives an inwardly directed magnetic pressure on the body. Such is the situation of a superconductor in a magnetic field. During the transition to the normal state this magnetic pressure does work on the superconduction electrons of the material, so changing the free energy of the superconductor.

It may appear that, since the bulk of the superconductor is shielded from the magnetic field by the superficial shielding currents, the electrons there should not be influenced by the field and their free energy should not be different from that in the absence of the field. This in fact is so until the occurrence of the superconducting-to-normal transition, during which the magnetic field penetrates. During this penetration, however, superconducting electrons at successively greater depths from the surface are accelerated by the penetrating magnetic field to reach a limiting velocity, at which, for a reason to be seen later, they can no longer carry a superconduction current. Thus the s-n transition involves simply an increase in the energy of the superconducting electrons to a level enjoyed by the electrons in the normal metal.

Thus, the magnetisation is a potential source of energy whose presence *per se* raises the free energy of the superconductor although the magnetic field does work on the conduction electrons only during the transition. This work is conserved by the electrons and does not increase the temperature of the superconductor; indeed in adiabatic conditions the temperature of a superconductor decreases during a magnetically induced transition to the normal state.

The increase in free energy of a superconductor due to its magnetisation is given by

$$\Delta G_s = \int_0^H \mu_0 M \, dH = \tfrac{1}{2}\mu_0 H^2 \ . \tag{5.1.2}$$

At the critical field H_c, the increase in free energy is obviously $\tfrac{1}{2}\mu_0 H_c^2$. This must therefore be the energy difference between the superconducting and normal states at the given temperature T in the absence of any field, i.e.

$$G_n = G_s + \tfrac{1}{2}\mu_0 H_c^2 \ , \tag{5.1.3}$$

where G_n and G_s are the free energies of the normal and superconducting phases respectively.

Now, entropy is related to free energy by the expression

$$S = -\frac{dG}{dT} \ . \tag{5.1.4}$$

Therefore the entropy difference between the superconducting and normal states, again in the absence of a field, is given by

$$S_n - S_s = -\mu_0 H_c \frac{dH_c}{dT} \ . \tag{5.1.5}$$

Further differentiation gives the difference in specific heat between the two phases because

$$C = T\frac{dS}{dT} \ . \tag{5.1.6}$$

Thus

$$C_n - C_s = \mu_0 \left[T H_c \frac{d^2 H_c}{dT^2} + T\left(\frac{dH_c}{dT}\right)^2 \right] \ . \tag{5.1.7}$$

Also, latent heat is related to entropy difference by the relationship

$$Q = T(S_n - S_s) = \mu_0 T H_c \frac{dH_c}{dT} \, . \tag{5.1.8}$$

From Equation (5.1.8) it is seen that in the presence of a magnetic field a latent heat is associated with the transition. On passing from the superconducting to the normal state under the influence of an increasing field, a superconductor absorbs heat.

From Equations (5.1.5) to (5.1.8) some important conclusions can be drawn.

From Equation (5.1.5) it is seen that, at the critical temperature T_c, the entropy difference between the superconducting and normal phases is zero because, at T_c, $H_c = 0$. Furthermore, the absence of any entropy change at this particular transition implies an absence of latent heat. This follows from Equation (5.1.8). From Figure 2.2.1 it is seen that dH_c/dT is negative, i.e. the critical field decreases with rising temperature. Consequently Equation (5.1.5) shows that the entropy of the normal state is greater than the entropy of the superconducting state. Since entropy is a measure of the disorder of a system (the entropy of the pieces increases as a game of chess proceeds), it is a conclusion that the superconducting state is one of greater 'order' than the normal state. It is as if in the superconducting state a portion of the electrons normally available to take part in random processes, such as the property of specific heat, were withdrawn and isolated in a low-energy state, from which they can be raised only by the supply of a finite amount of energy, such as the energy of magnetisation.

The 'degree or order' depends on the temperature, as will be seen later. Physically, the parameter which is ordered is the motion of the conduction electrons.

It is a basic precept of thermodynamics that at absolute zero the entropy of all systems is zero. Thus the entropy difference between the two phases is zero at $T = 0$. The form of the entropy–temperature curve for the two states is therefore as shown in Figure 5.1.1.

By reference to Figure 5.1.1 it can be inferred that the difference in specific heats, $C_n - C_s$, is negative near T_c but that $C_n - C_s$ changes sign at a lower temperature. Figure 5.1.2 shows the form of the two specific heat curves. A specific form for Equation (5.1.7) occurs at the critical temperature T_c, where $H_c = 0$. There,

$$C_n - C_s = -\mu_0 T_c \left(\frac{dH_c}{dT}\right)^2 \, . \tag{5.1.9}$$

To complete this brief statement of the thermodynamics of the superconducting to normal transition the form of the curve of critical field H_c versus temperature will be derived from the experimentally observed forms of the specific heat curves. From experimental data expressions for C_n and C_s are

$$C_n = \gamma T + \alpha T^3$$
$$C_s = \beta T^3 \quad \text{(approximately)}.$$

It must be stressed that these are approximate functions obtained from experimental observation. They are empirical relationships.

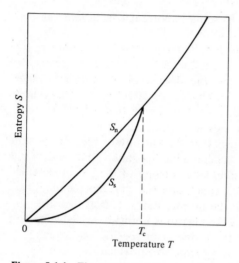

Figure 5.1.1. The entropy of the normal and superconducting states of a superconductor as a function of temperature.

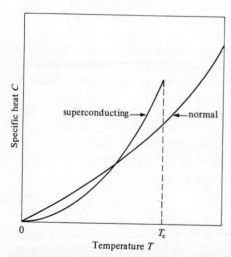

Figure 5.1.2. The specific heat of the normal and superconducting states of a superconductor as a function of temperature.

The linear term in temperature in the expression for C_n is intimately associated with an electronic component whilst the cubic term is the lattice specific heat. The absence of the electronic specific heat term from the expression for C_s does not, however, imply that no electrons whatever are available for random processes in the superconducting state. The inference is only that the fraction available for random processes increases with temperature in such a way as to give a net cubic temperature dependence for electronic specific heat in the superconducting phase.

From Equation (5.1.6)

$$S = \int \frac{C}{T}\,\mathrm{d}T$$

so

$$S_n - S_s = \gamma T + \tfrac{1}{3}\alpha T^3 - \tfrac{1}{3}\beta T^3 + K .$$

But, at $T = 0$, $S = 0$ and $S_n - S_s = 0$; therefore $K = 0$. At T_c, $S_n - S_s = 0$, so that

$$\gamma T_c + \tfrac{1}{3}(\alpha - \beta)T_c^3 = 0 \qquad\qquad (5.1.10)$$

and, at temperatures below T_c,

$$S_n - S_s = \gamma T - \frac{\gamma T^3}{T_c^2} . \qquad\qquad (5.1.11)$$

From Equation (5.1.5)

$$S_n - S_s = -\mu_0 H_c \frac{\mathrm{d}H_c}{\mathrm{d}T} ,$$

whence by integration

$$\tfrac{1}{2}\mu_0 H_c^2 = \gamma(\tfrac{1}{4}T^4 T_c^{-2} - \tfrac{1}{2}T^2) + N . \qquad\qquad (5.1.12)$$

At $T = T_c$, $H_c = 0$ so that $N = \tfrac{1}{4}\gamma T_c^2$ and, at $T = 0$, $H_c = H_0$, the critical field at absolute zero, so that

$$\gamma = 2\mu_0 H_0^2 T_c^{-2} .$$

Substituting for γ and N in Equation (5.1.12) gives

$$H_c = H_0 \left[1 - \left(\frac{T}{T_c}\right)^2\right] . \qquad\qquad (5.1.13)$$

This parabolic variation of critical field with temperature holds closely, although not exactly, for most superconductors and demonstrates the validity of the thermodynamic treatment. Typical curves of critical field as a function of temperature are shown in Figure 2.2.1 for a number of soft superconductors.

Thus we have derived a thermodynamic connection between two observable phenomena, the variation of critical field with temperature and the differences in specific heat of the superconducting and normal phases.

We tacitly assumed, however, that the $s-n$ transition is reversible, for that is implied in the relationships used to obtain entropy and specific heat from the free energy. Indeed the phenomenological thermodynamic theory was developed at a time when it was thought that the transition was not reversible.

However, in 1933 Meissner and Ochsenfeld discovered that, when a pure crystal of tin was cooled in a magnetic field the field was almost completely expelled from the crystal. This effect, known as the Meissner effect, occurs only in pure soft superconductors and it is conclusive proof of the reversibility of the superconducting normal transition, at least in such materials. It also implies that the fundamental property of soft superconductors is not perfect conductivity but complete diamagnetism.

It will in fact be shown later that other types of superconductor show little or no Meissner effect. The transition in these superconductors is nevertheless truly reversible and indeed thermodynamic reversibility is a basic, if sometimes obscure, property of all superconductors.

5.2 Electrodynamics and the London equations†

If diamagnetism is to exist in a superconductor the net field strength within it must be less than the field strength outside: indeed, for complete diamagnetism, $M = -H$ and the field within is zero.

To maintain this difference in field strengths a surface current must flow. Since it is unreasonable to suppose that the current density within a superconductor is infinite, the field strength must fall from its external level to a very small value at a finite rate. The distance for a 63% decrease in field strength is in fact about 10^{-5} cm in soft superconductors and this distance is called the penetration depth. It can be calculated simply as follows.

Consider the penetration of an external field into a semi-infinite superconductor. Let n_s be the number density of superconducting electrons. Let m be the mass of an electron. Let e be its charge. Then the current density is given by

$$J = n_s e v ,$$

$$(5.2.1)$$

where v is the velocity.

Further

$$m \frac{dv}{dt} = eE ,$$

$$(5.2.2)$$

where E is a voltage gradient.

Eliminating v between Equations (5.2.1) and (5.2.2) gives

$$\frac{dJ}{dt} = \frac{n_s e^2}{m} E .$$

$$(5.2.3)$$

† London (1950, p.27 ff.).

Now, to relate J and E to H it is necessary only to note the one-dimensional statement of Maxwell's equations, namely

$$J = -\frac{dH}{dx}, \qquad \frac{dE}{dx} = -\frac{dB}{dt}, \qquad B = \mu_0 H.$$

(Here we note that, within the penetration region, the local field strength H is equal to the local flux density B in unrationalised units.)

With these equations, expression (5.1.3) becomes

$$\lambda^2 \frac{d^2\dot{H}}{dt} = \mu_0 \dot{H}, \tag{5.2.4}$$

where

$$\lambda^2 = \frac{m}{\mu_0 n_s e^2} \tag{5.2.5}$$

and $\dot{H}$ denotes dH/dt.

The solution to Equation (5.2.4) which satisfies the condition that the field strength deep inside the superconductor is zero is

$$\dot{H}(x) = \dot{H}_e \exp\left(-\frac{x}{\lambda}\right), \tag{5.2.6}$$

where $\dot{H}_e$ is the time-varying field at the surface.

However, Equation (5.2.6) is not the full description of a superconductor although it describes the behaviour of a perfect conductor. This is concluded because in Equation (5.2.6), for $x \gg \lambda$, $\dot{H}(x) = 0$, i.e. the field cannot change deep within the conductor. The Meissner effect contradicts this.

Integrating Equation (5.2.4) with respect to time yields

$$\lambda^2 \frac{d^2(H - H_m)}{dx^2} = H - H_m, \tag{5.2.7}$$

where H_m is an arbitrary constant. H_m should in fact be the field in the conductor when it became superconducting.

However, in order to reconcile this theoretical treatment with the Meissner effect, H_m must be set equal to zero. Then Equation (5.2.7) is solved as

$$H(x) = H_e \exp\left(-\frac{x}{\lambda}\right), \tag{5.2.8}$$

which is the equation of flux penetration proposed by F. and H. London in 1935. The penetration of flux into a thick superconductor is shown in Figure 5.2.1.

When $x = \lambda$, the field strength is 36% of the surface value and this distance is called the penetration depth. For lead, for instance, taking n_s, the superconducting electron density, to correspond to two electrons per atom and m and e with their usual values, $\lambda = 0.5 \times 10^{-6}$ cm.

(Experimental values of λ for various superconductors are some five to ten times greater than the theoretical value for reasons discussed later.) Soft superconductors substantially larger than this typical size will exhibit almost complete diamagnetism. The maximum current density at the surface of a soft superconductor can be found from Equation (5.2.8) since

$$J_{max} = -\frac{dH}{dx_{max}} = -\frac{H_c}{\lambda} .$$ (5.2.9)

Typically $H_c = 400$ A cm^{-1} and $\lambda = 0\cdot5 \times 10^{-5}$ cm experimentally, so $J_{max} = 8 \times 10^7$ A cm^{-2}.

It must be emphasised here that the London model of a soft super-conductor is essentially local. The interaction of the magnetic flux with the superconducting electrons is assumed to be purely local. In fact, however, because of the range of coherence of the superconducting electron pairs the effect of a magnetic field at one point is manifest over a large volume of the superconductor. The mathematical treatment of non-local electrodynamics is too complex to be considered here; however, in Section 7.3 we analyse the rather remarkable influence which the coherence length has on the behaviour of a superconductor.

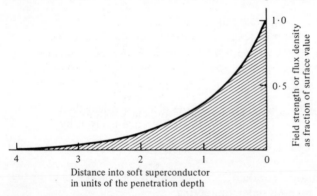

Figure 5.2.1. The penetration of magnetic flux into a semi-infinite slab of a soft superconductor.

5.3 The two-fluid model of superconductivity†

The foregoing electrodynamic treatment relates the electronic properties of a superconductor to its magnetic properties in classical terms. We have also seen how its magnetic property, that is its critical field strength, varies with temperature. We could therefore conclude that there exists a simple phenomenological dependence of the electronic properties of a super-conductor on its temperature. Such a dependence was indeed derived by Gorter and Casimir in 1934 by a two-fluid analogue of the treatment of superfluidity in helium.

† Gorter and Casimir (1934); London (1950, p.83 ff.); Schoenberg (1952, p.194 ff.).

In the two-fluid model of superconductivity it is assumed that a fraction $1 - \omega$ of the electrons available for conduction remain in the normal state and that a fraction ω has 'condensed' into a superconducting state of lower energy. Thus ω is the degree of order referred to in Section 5.1. The fraction ω is a function of temperature yet to be determined, but, at $T = 0$, $\omega = 1$ and, at $T = T_c$, $\omega = 0$. Since ω represents the degree of superconducting order, it is frequently described as the order parameter and is referred to again in Section 7.1.

If we consider the energy of the electrons only and ignore the contribution of the lattice, which is reasonably assumed to be the same in both the normal and superconducting states, the free energy per unit volume of the superconductor is given by

$$\frac{G_s}{N} = f_n(1 - \omega)g_n(T) + f_s(\omega)g_s(T) \; , \tag{5.3.1}$$

where N is the total number of superconducting and normal electrons per unit volume, $g_n(T)$ is the temperature-dependent free energy of a normal electron, $g_s(T)$ is that of a superconducting electron, and $f_n(1 - \omega)$, $f_s(\omega)$ are functions of ω yet to be determined. The free energy of a normal electron is taken as

$$g_n(T) = -\frac{\gamma T^2}{2N} \cdot \tag{5.3.2}$$

[The familiar form of the normal electronic specific heat,

$$C_{en} = \gamma T \; ,$$

is derived from this using Equations (5.1.4) and (5.1.6).]

Now the free energy of a superconducting electron is less than that of the normal electron. At absolute zero this difference is given by

$$g_s(0) - g_n(0) = -\frac{\mu_0 H_0^2}{2N} \; ,$$

but, as $g_n(0) = 0$, then

$$g_s(0) = -\frac{\mu_0 H_0^2}{2N} \cdot \tag{5.3.3}$$

The choice of the functions f_n, f_s must satisfy experimentally observed characteristics, such as the temperature variation of specific heat. This was previously taken in Section 5.1 as $\gamma T + \alpha T^3$ in the normal state and as βT^3 in the superconducting state.

The functions which eventually satisfy these criteria are

$$\left. \begin{array}{l} f_s(\omega) = \omega \\ f_n(1 - \omega) = (1 - \omega)^{\frac{1}{2}} \; . \end{array} \right\} \tag{5.3.4}$$

Substitution of these functions into Equation (5.3.1), together with the expression for $g_n(T)$ and $g_s(T)$ from Equations (5.3.2) and (5.3.3) gives

$$G_s = -\tfrac{1}{2}(1 - \omega)^{\frac{1}{2}}\gamma T^2 - \tfrac{1}{2}\omega\mu_0 H_0^2 \; , \tag{5.3.5}$$

where G_s is the total free energy of the superconductor due to normal and superconductive electrons. Now, if the proportions of the types of electron, i.e. if the two phases, are in equilibrium, their total free energy will remain constant against a small change in the proportions of the two types. (If this were not so, the proportion would change in order to obtain the lowest free energy for the whole system.)

The condition that the rate of change of free energy be zero is a commonly used criterion for the equilibrium of a thermodynamic system. Thus, by differentiating Equation (5.3.5) and equating it to zero,

$$1 - \omega = \left(\frac{\gamma T^2}{2\mu_0 H_0^2}\right)^2$$

is the condition for equilibrium. At $T = T_c$, $\omega = 0$, so that

$$\gamma = 2\mu_0 H_0^2 T_c^{-2}$$

(as in Section 5.1) and

$$\omega = 1 - \left(\frac{T}{T_c}\right)^4 . \tag{5.3.6}$$

If we substitute into Equation (5.3.5) for ω and H_0,

$$G_s = -\tfrac{1}{4}\gamma T^4 T_0^{-2} - \tfrac{1}{4}\gamma T_c^2 . \tag{5.3.7}$$

Using Equations (5.1.4) and (5.3.5), we obtain the entropy of the electrons in the superconducting phase:

$$S_{es} = -\frac{dG_s}{dT} = \gamma T^3 T_c^{-2} . \tag{5.3.8}$$

Again using Equation (5.1.6), we find the electronic specific heat of the electrons in the superconducting phase:

$$C_{es} = 3\gamma T^3 T_c^{-2} . \tag{5.3.9}$$

Since the lattice specific heat is also of the form αT^3, the total specific heat of the superconducting phase is $C_s = \beta T^3$. This confirms that the functions $f_n(1 - \omega)$ and $f_s(\omega)$ were chosen correctly.

The two-fluid model can be used to predict the temperature variation of the London penetration depth.

It was seen in Equation (5.2.5) that

$$\lambda^2 = \frac{m}{\mu_0 n_s e^2} .$$

Now we write

$$n_s = \omega N = N\left[1 - \left(\frac{T}{T_c}\right)^4\right] \tag{5.3.10}$$

and put

$$\lambda_0^2 = \frac{m}{\mu_0 N e^2} \ ,$$

where λ_0 is the penetration depth at absolute zero. Then

$$\frac{\lambda}{\lambda_0} = \left[1 - \left(\frac{T}{T_c} \right)^4 \right]^{-\frac{1}{2}} . \tag{5.3.11}$$

This expression is plotted in Figure 5.3.1. It is in close agreement with experiment, which is good proof that the two-fluid model and the order parameter are appropriate concepts which adequately describe some of the phenomenological aspects of the behaviour of superconductors.

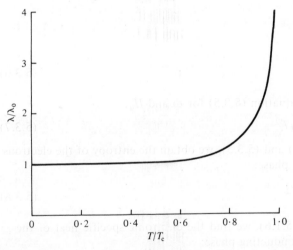

Figure 5.3.1. The variation of the penetration depth of a soft superconductor as a function of the reduced temperature T/T_c.

5.4 High-frequency effects†

It was seen in Section 5.2 that the effect of a voltage gradient in a super-conductor is to accelerate the superconducting electrons. Because the voltage gradient is caused by the movement of flux in the superconductor, the acceleration of the superconducting electrons continues until a current flow has been produced which inhibits further flux movement.

During the period of acceleration of the superconducting electrons, the voltage gradient can also act upon and accelerate 'normal' electrons, thereby causing transient dissipative current flow.

If therefore a high-frequency alternating voltage gradient is applied to a superconductor, the superconducting electrons will be in a state of almost

† Gorter and Casimir (1934); London (1950, p.83 ff.); Schoenberg (1952, p.194 ff.).

continuous acceleration in alternate directions. Under these conditions the voltage gradient will affect the normal electrons causing a continuous dissipative current flow. When the superconducting and normal current densities are comparable, it is to be expected that 'normal' properties will return to the superconductor. It is a simple calculation to determine the frequency at which this should happen.

Suppose a voltage gradient varying sinusoidally in time is applied to the superconductor, such that $E = E_0 \sin \omega t$. Let the normal and super-conducting current densities be j_s and j_n respectively and let ρ_n be the normal state resistivity. Then, from Equation (5.1.3), by integration,

$$j_s = \frac{n_s e^2}{m \omega} E_0 \cos \omega t \ . \tag{5.4.1}$$

At a temperature above absolute zero there will be present 'normal' electrons which can take part in dissipative processes. The normal current density will be

$$j_n = \frac{n_n}{N \rho} E_0 \sin \omega t \ , \tag{5.4.2}$$

where n_n is the density of normal electrons, ρ is the normal-state resistivity, and N is the total conduction electron density. j_s and j_n are instantaneous values of current density. The corresponding r.m.s. values are

$$J_s = \frac{n_s e^2}{m \omega} \frac{E_0}{(2)^{1/2}} \tag{5.4.3}$$

and

$$J_n = \frac{n_n}{N \rho} \frac{E_0}{(2)^{1/2}} \ . \tag{5.4.4}$$

Thus

$$\frac{J_n}{J_s} = \frac{m \omega n_n}{n_s e^2 N \rho} \ . \tag{5.4.5}$$

If we substitute for $m/n_s e^2$ from Equation (5.2.5),

$$\frac{J_n}{J_s} = \frac{\mu_0 \lambda^2 \omega n_n}{N \rho} \ . \tag{5.4.6}$$

Now λ and n_n/N can be written in terms of temperature using the two-fluid model. Thus, with Equations (5.3.10) and (5.3.11),

$$\frac{J_n}{J_s} = \frac{\mu_0 \lambda_0^2 \omega}{\rho} \frac{(T/T_c)^4}{1 - (T/T_c)^4} \ , \tag{5.4.7}$$

where λ_0^2 is the penetration depth at absolute zero.

By way of example the frequency at which significant resistive effects set in can be calculated. For lead,

$\lambda_0 = 0 \cdot 39 \times 10^{-5}$ cm (experimental),
$T_c = 7 \cdot 175°$K,
$\rho = 10^{-8}$ Ω cm.

The criterion for the appearance of resistance will be taken as

$|J_n| = |J_s|$.

Then, if an operating temperature of $2 \cdot 0°$K is assumed,

$\omega = 40 \times 10^{10}$ rad s^{-1}

and

$f = 63500$ MHz.

This frequency is in the infrared region. It is high enough to allow resonant circuits of very high magnification (Q) to be constructed for linear accelerators and filters (see Section 16).

Although the frequency at which resistive effects commence increases as the temperature is lowered, a limit is reached at about 100 GHz at which frequency quantum effects produce a rapid increase in resistance. This is explained in Section 3.1.

Quantum effects

6.1 Flux quantisation†

The property of zero resistance in a superconductor implies a conservation of energy that can be satisfied only by quantisation.

Consider what would befall a current set up in a ring of a normal conductor of zero resistance. Because the electrons must continually accelerate centripetally they would emit cyclotron radiation, a phenomenon occurring whenever free electrons are accelerated. In contrast a current set up in a superconducting ring will persist indefinitely losing no energy whatever, as long as the superconductor is maintained below its critical temperature. The very fact that there is no energy loss from a persistent superconducting current demands the quantisation of the energy of the superconducting electron flow. Just as the energy levels of the electrons in orbit around a proton in a hydrogen atom are quantised, so are the energy levels of superconducting persistent current. Let it be stressed, however, that this quantisation applies only to wholly superconducting circuits, containing no resistive elements and no sources of e.m.f. As a consequence of this quantisation of superconducting electronic energy levels, the flux in a closed superconducting loop is also quantised.

The starting point for the development of the theory of flux quantisation is the expression for the momentum of an electron in a magnetic field, namely

$$p = mv + eA , \qquad (6.1.1)$$

where p is the canonical momentum, m is the electron mass, v is the electron velocity, e is the electron charge, and A is the magnetic vector potential.

In a circular circuit,

$$2\pi r A = \phi , \qquad (6.1.2)$$

where r is the radius of the circuit and ϕ is the flux contained in it.

The origin of Equation (6.1.1) may not be immediately obvious. However, it is easily derived.

The total energy stored in an inductive loop is

$$\epsilon = \tfrac{1}{2}LI^2 + \tfrac{1}{2}nmv^2 , \qquad (6.1.3)$$

where n is the number of electrons in the loop and the other symbols have their usual meaning. In most practical electrical circuits, the first term predominates and need alone be considered. However, in circuits of very small dimension the first term may be smaller than the second.

† Weisskopf (1962).

The current may be expressed as

$$I = \frac{nve}{2\pi r} \; ; \tag{6.1.4}$$

substituting for I in Equation (6.1.3) gives

$$\epsilon = \tfrac{1}{2} L \left(\frac{nve}{2\pi r} \right)^2 + \tfrac{1}{2} nmv^2 . \tag{6.1.5}$$

If this energy is differentiated with respect to velocity, the momentum is obtained; thus,

$$p = Lv \frac{ne}{2\pi r} + nmv . \tag{6.1.6}$$

The inductance L, which is a constant, may be replaced by

$$L = \frac{\phi}{I} = \frac{2\pi r \phi}{nve} . \tag{6.1.7}$$

Substituting this into Equation (6.1.6) gives

$$p = nmv + \frac{ne\phi}{2\pi r}$$

$$= nmv + neA . \tag{6.1.8}$$

We can now use this expression for the canonical momentum of electrons in a magnetic field to derive the magnitude of the flux quantum in a superconducting loop.

We start by considering a superconducting cylinder, the inner radius and wall thickness of which are much greater than the superconducting penetration depth. Under these conditions the contribution of the term mv in the momentum equation can be ignored, just as its integral, the kinetic energy, can be ignored in macroscopic electrical circuits.

Then we have for the momentum

$$p = neA , \tag{6.1.9}$$

As mentioned in Section 3.1, superconducting electrons are paired and we considered the momentum of a pair, so that n is 2.

Now it is a fundamental tenet of quantum physics that associated with a particle of momentum p is a wavelength λ, such that

$$p = \frac{h}{\lambda} , \tag{6.1.10}$$

where h is Planck's constant.

What λ implies, in terms as near to a classical model as can be approximated, is this. A moving particle has a wavefunction which expresses the probability that the particle will be at a particular point in space at a particular time. λ is then the distance between points of equal probability at a given time.

If an electron is moving in a closed orbit such that the path length of one orbit is an integral number of wavelengths, then the orbit is quantised and the electron neither gains nor loses energy. The wavefunction of a quantised orbit is a standing wave in which the probability of finding the electron at a particular point does not vary in time. It is not a difficult exercise for the imagination to see that the implication of the constancy of the probability at a point in space is that no energy is gained or lost by the electron to disturb its spatial distribution. Thus, for a particle in an orbit of radius r, the condition for quantisation is

$$2\pi r = N\lambda \,, \tag{6.1.11}$$

where N is any integer.

In the case of the superconducting cylinder, the condition that a super-conduction current should flow perpetually without energy loss is then

$$2\pi r = \frac{Nh}{p} \,. \tag{6.1.12}$$

Now,

$$p = neA \ (= 2eA \text{ for an electron pair}),$$

so that

$$2\pi r = \frac{Nh}{2eA} \,. \tag{6.1.13}$$

But

$$2\pi rA = \phi \,,$$

where ϕ is the flux enclosed in the cylinder; therefore, for a lossless circulating current,

$$\phi = \frac{Nh}{2e} \,. \tag{6.1.14}$$

This result is of outstanding importance. It means that in a closed super-conducting circuit the flux is quantised in units of $h/2e$, $2 \cdot 07 \times 10^{-15}$ Wb. This flux quantum is given the symbol ϕ_0.

The magnitude of ϕ_0 has been confirmed experimentally by several workers (Doll and Nabauer, 1961; Deaver and Fairbank, 1961). One experiment involved the measurement of the torque exerted by a transverse field on a tube containing trapped flux. In this experiment a very small superconducting cylinder was formed by evaporating lead onto a $0 \cdot 6$ mm length of quartz fibre, 10 μm in diameter as shown in Figure 6.1.1. The quartz fibre was suspended at the middle of the lead film by a torsion thread which carried a small mirror. With neither trapped field nor external field, the natural damping constant of the system was measured. The fibre and film were then cooled in a magnetic field parallel to the axis of the fibre. This field was then removed and an intermittent field applied at right angles. By switching this measuring field at regular intervals a steady oscillation of the fibre with its lead film and trapped field could be maintained. From a knowledge of the natural damping coefficient of the system, the measuring field, and the dimensions of the lead cylinder, the trapped flux was inferred in terms of resonance amplitude and measuring field. In this way the trapped flux was measured as a function of the trapping field, which was varied from 0 to 0.4 A cm^{-1} in very small steps. The results are shown in Figure 6.1.2. It is quite clear that the trapped flux only takes on values of multiples of 2.07×10^{-15} Wb. It will also be noticed, however, that the field strength in the cylinder may exceed or may be less than the initial external field by up to one-half of the flux quantum. Thus it is to be inferred that the points A in Figure 6.1.2 represent a state of lower energy than do the points B. How this occurs and how it is exploited in a device is considered in Section 12.

The assumption was made earlier that, if the bore and wall thickness of a cylinder were large compared with the penetration depth, the kinematic term mv in the canonical momentum could be ignored. We shall now consider the case of a thin-walled superconducting cylinder of small bore: here we must consider both terms in the canonical momentum. Figure 6.1.3 shows a section through the cylinder. The magnetic flux within the dotted circle is denoted ϕ_r and the current density at the radius r is denoted j_r. The radius of the dotted line is r. If the electron flow along the dotted path is to be a persistent quantum state, then the

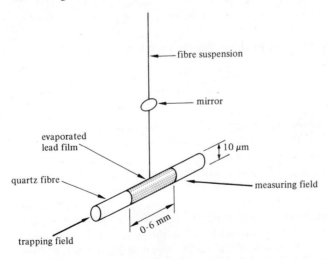

Figure 6.1.1. Apparatus used in the measurement of the magnitude of the flux quantum.

expression for the canonical momentum must be quantised; that is, as before,

$$p = mv + eA \qquad \frac{h}{\lambda} = \frac{Nh}{2\pi r} \; .$$

Thus

$$mv_r + \frac{e\phi}{2\pi r} = \frac{Nh}{2\pi r} \; . \qquad\qquad (6.1.15)$$

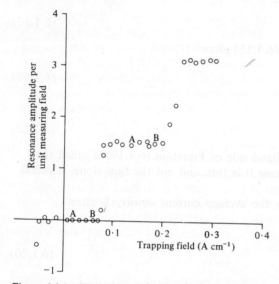

Figure 6.1.2. The variation of the trapped flux as a function of field strength. The ordinates are in fact the amplitude of the angular swing divided by the measuring field strength and are proportional to the trapped flux.

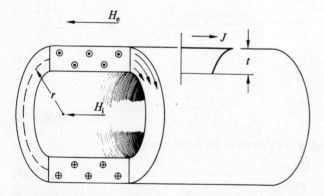

Figure 6.1.3. The penetration of flux into a thin-walled cylinder of a soft superconductor and the distribution of currents in it. Although the current density varies through the cylinder wall as shown in the curve, it is assumed to be constant because t is less than the penetration depth.

v_r can be replaced by the current density as

$$v_r = \frac{j}{n_s e} , \tag{6.1.16}$$

where n_s is the density of superconducting electrons and e is the electronic charge.

Using Equation (5.2.5) for the superconducting penetration depth λ, we can put

$$mv_r = \mu_0 \lambda^2 je . \tag{6.1.17}$$

Substituting into Equation (6.1.15) gives

$$\mu_0 \lambda^2 je + \frac{e\phi_r}{2\pi r} = \frac{Nh}{2e} , \tag{6.1.18}$$

which reduces to

$$2\pi r \mu_0 \lambda^2 j + \phi_r = N\phi_0 . \tag{6.1.19}$$

The expression on the left-hand side of Equation (6.1.19) is called the fluxoid and in the general case it is this, and not the flux alone, which is quantised.

In a thin-walled cylinder the average current density is given approximately by

$$j = \frac{I}{t} , \tag{6.1.20}$$

where I is the current per unit length and t is the wall thickness. Furthermore, if the external field strength is zero, then the field strength within the cylinder is given by

$$H_i = I , \tag{6.1.21}$$

so that

$$j = \frac{H_i}{t} . \tag{6.1.22}$$

Also

$$\phi_r = \mu_0 \pi r^2 H_i . \tag{6.1.23}$$

Substituting for j and ϕ_r in Equation (6.1.19), we obtain

$$\frac{2\pi r \lambda^2 H_i}{t} + \pi r^2 H_i = \frac{N\phi_0}{\mu_0} , \tag{6.1.24}$$

whence

$$\pi r^2 H_i = \frac{N\phi_0}{\mu_0} \left(1 + \frac{2\lambda^2}{rt}\right)^{-1} . \tag{6.1.25}$$

Thus the flux trapped by a cylinder is less than an integral number of quanta. However, except for cylinders of bore and thickness comparable with the penetration depth λ, the difference is very small.

Suppose now that the superconducting cylinder is in an external magnetic field H_e and has an internal magnetic field H_i. Then the current flowing in the wall of the cylinder must satisfy the condition

$$jt = \pm(H_e - H_i) , \tag{6.1.26}$$

in which we assume the current density to be roughly uniform (that is $t < \lambda$). The duality of sign allows for either an excess or a deficit of internal field strength (trapping or shielding). If we so substitute for j in Equation (6.1.18) and put $\phi_r = \mu_0 \pi r^2 H_i$, we obtain

$$H_i \left(\pi r^2 - \frac{2\pi\lambda^2 r}{t} \right) + H_e \frac{2\pi\lambda^2 r}{t} = \frac{N\phi_0}{\mu_0} , \tag{6.1.27}$$

which reduces to

$$\pi r^2 H_i \left(1 - \frac{2\lambda^2}{rt} \right) = \frac{N\phi_0}{\mu_0} - H_e \frac{2\pi\lambda^2 r}{t} . \tag{6.1.28}$$

The energy stored in the cylinder is given by

$$\epsilon = \tfrac{1}{2} L I^2 ,$$

where L is the inductance and I the current flowing in the wall. The current flowing in the wall is given by

$$H_e - H_i = \frac{I}{l} , \tag{6.1.29}$$

where l is the length of the cylinder. Thus

$$\epsilon = \tfrac{1}{2} L(H_e - H_i)^2 l . \tag{6.1.30}$$

Substituting for H_i from Equation (6.1.28), we find

$$\epsilon = \tfrac{1}{2} L l \left[H_e - \frac{N\phi_0/\mu_0 - H_e(2\pi\lambda^2 r/t)}{\pi r^2 (1 - 2\lambda^2/rt)} \right]^2 , \tag{6.1.31}$$

which reduces to

$$\epsilon = C \left(\pi r^2 H_e - \frac{N\phi_0}{\mu_0} \right)^2 . \tag{6.1.32}$$

C is a constant which includes all the geometric parameters. The variation of ϵ as a function of the external field strength H_e is shown in Figure 6.1.4. Each parabolic curve corresponds to an integral number of flux quanta: for a given level of external field strength, the energy of the cylinder can take on one of a series of particular values, one of which will be a minimum.

Let us consider the curves corresponding respectively to N and $N+1$ flux quanta within the cylinder. At a particular value of H_e the energies will be the same. This value of field strength is given by

$$2H_e = \frac{(2N+1)\phi_0}{\pi r^2 \mu_0} \tag{6.1.33}$$

If the external field strength rises slightly above the level given by this condition, a tendency develops for one flux quantum to enter the cylinder, thereby lowering the energy. If H_e decreases, one flux quantum will tend to leave the cylinder. The point at which either of these processes occurs depends on the potential change in energy of the system. A flux quantum will move through the cylinder wall when the change in energy resulting from this movement equals the energy required to form a fluxoid in the cylinder wall. Thus, if the external field is steadily increased, a series of energy changes, and therefore of flux movements, such as are represented by the points 1, 2 in Figure 6.1.4, will result. During a steady decrease of external field strength a similar series of flux changes occurs.

This discontinuous movement of flux is exploited in the quantum magnetometer and is described further in Section 12.

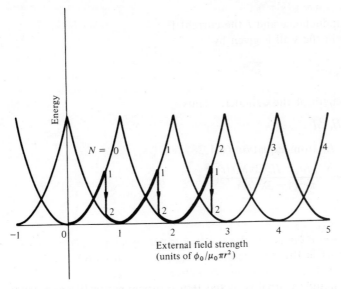

Figure 6.1.4. The variation of the energy of a superconducting cylinder as a function of the strength of an external field. The points 1, 2 represent typical abrupt changes in the energy of the cylinder as one quantum of flux enters.

6.2 The London moment†

To complete this section on flux quantisation we shall derive the very simple expression for the London moment of a rotating superconductor.

In the expression for the canonical momentum of a superconducting electron pair,

$$p = 2mv + 2eA ,$$ (6.2.1)

the velocity term represents the movement of the electron pairs through the lattice of the superconductor. Suppose a simply connected super-conductor traps a flux ϕ so that, to satisfy the condition of quantisation,

$$2\pi r \frac{mv}{e} + \phi = N\phi_0 .$$ (6.2.2)

If the superconductor is now rotated at an angular speed ω rad s^{-1}, the velocity of the electron pairs changes by an amount ωr. Thus, the momentum of an electron pair changes by $2m\omega r$ and the first term of Equation (6.2.2) changes by

$$2\pi r \frac{2m\omega r}{2e} .$$

In order to retain the necessary quantum condition the flux must change. This change is given by

$$\Delta\phi + 2\pi r \frac{m\omega r}{e} = N\phi_0 .$$ (6.2.3)

In the simplest case $N = 0$, so that

$$\Delta\phi = -2\pi r \frac{m\omega r}{e} .$$ (6.2.4)

Now

$$\Delta\phi = \mu_0 \Delta H \pi r^2 ,$$ (6.2.5)

where ΔH is the change in field strength, so that

$$\Delta H = -\frac{2m\omega}{\mu_0 e} .$$ (6.2.6)

ΔH is the London moment of the rotating superconductor. It is one of the causes of drift in the superconducting gyroscope, discussed in Section 16.2.

By way of example it may be noted that a superconductor, spinning at 1000 rev s^{-1}, generates a London moment of $1 \cdot 14$ mA cm^{-1}, a small effect in practical terms.

---

† London (1950, p.78 ff.); Bol and Fairbank (1964).

The theory of type II superconductors

7.1 The critical field of thin superconductors

In Section 5.2 we saw that the field strength in a soft superconductor falls exponentially from its surface value with a characteristic length λ, the penetration depth, found experimentally to be about 10^{-5} cm.

Suppose a soft superconductor is in the form of a plate of thickness less than the penetration depth. The penetration profile will then be roughly as shown in Figure 7.1.1. By using Equation (5.1.4), together with the condition that $H = H_e$ at $x = 0$ and that $dH/dx = 0$ in the middle of the plate, the expression for the variation of the internal field strength (or flux density) can be shown to be

$$H(x) = \frac{B(x)}{\mu_0} = H_e \frac{\cosh[(a-x)/\lambda]}{\cosh(a/\lambda)} \; , \tag{7.1.1}$$

where a is the semithickness of the plate, x is the depth below the surface, and H_e is the external field. Now, from Equation (5.1.2) we know that the energy of the superconducting phase is increased by having a magnetisation $-M$ and that this increase is given by

$$\Delta G_s = \int_0^H \mu_0 M \, dH \; .$$

The average magnetisation of the thin plate is given by

$$M = \frac{1}{a} \int_0^a H \, dx - H_e \; , \tag{7.1.2}$$

which, by substitution of H from Equation (7.1.1) gives

$$M = -H_e \left[1 - \frac{\lambda}{a} \tanh\left(\frac{a}{\lambda}\right) \right] \; . \tag{7.1.3}$$

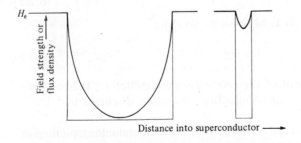

Figure 7.1.1. The profile of the penetration of magnetic flux into a thin plate of soft superconductor: for comparison, the penetration of flux into a thick superconductor. Note the difference in magnetisation.

† Schoenberg (1952, Appendix II, p.233).

Thus, the increase in energy of the thin plate is given by

$$\Delta G_s = \mu_0 \int_0^{H_e} \left\{ H_e \left[1 - \frac{\lambda}{a} \tanh\left(\frac{a}{\lambda}\right) \right] \right\} dH$$

$$= \tfrac{1}{2} \mu_0 H_e^2 \left[1 - \frac{\lambda}{a} \tanh\left(\frac{a}{\lambda}\right) \right] . \tag{7.1.4}$$

At transition to normality,

$$\Delta G_s = G_n - G_s = \tfrac{1}{2} \mu_0 H_c^2 .$$

Thus the critical field of the thin plate is given by

$$H_{ec}^2 \left[1 - \frac{\lambda}{a} \tanh\left(\frac{a}{\lambda}\right) \right] = H_c^2 ,$$

that is

$$\frac{H_{ec}}{H_c} = \left[1 - \frac{\lambda}{a} \tanh\left(\frac{a}{\lambda}\right) \right]^{-\frac{1}{2}} , \tag{7.1.5}$$

where H_{ec} is the critical field for the thin plate and H_c is the bulk critical field. If $a \ll \lambda$, Equation (7.1.5) reduces to

$$\frac{H_{ec}}{H_c} \approx (3)^{\frac{1}{2}} \frac{\lambda}{a} . \tag{7.1.6}$$

Here then there appears to be a way of achieving high-field super-conductivity. For instance, lead, with a bulk critical field of 640 A cm^{-1} at 0°K and a penetration depth of $3 \cdot 9 \times 10^{-6}$ cm, will have a critical field of 5 kA cm^{-1} for films of thickness 10^{-6} cm.

7.2 Surface energy†

At this stage it is worth reviewing the influence of free energy on the superconducting to normal transition. In Section 2.2 we saw that the free energy difference between the superconducting and normal phases was equal to the energy of magnetisation per unit volume at the critical field strength ($G_n - G_s = \tfrac{1}{2} \mu_0 H_c^2$). Furthermore, we saw that this increase in free energy was imparted to the superconducting electrons layer by layer as the transition zone moved into the superconductor. Because the magnetisation of a macroscopic superconductor is uniform throughout its bulk it is easy to see how the increase in energy at transition is the same everywhere.

However, in the present example of a thin plate the magnetisation is not uniform and it would seem that the electrons at the centre of the thin plate would experience a greater increase in energy than those at the edges. There is apparently no constant transition region to move uniformly inwards. The answer to this paradox is obtained by considering the

† London (1950, p.125 ff.).

velocity of the electrons at the surface, for this is the direct cause of the breaking up of the electron pairs and hence of the quenching of super-conductivity.

Thus for a semi-infinite superconductor at the critical field the surface current density is given by

$$J_c = -\frac{dH}{dx} \tag{7.2.1}$$

which from Equation (5.2.9) is $-H_c/\lambda$. In the case of a thin super-conducting plate the surface current density is obtained by differentiating Equation (7.1.1). Thus

$$J_{tc} = \left(\frac{dH}{dx}\right)_{x=0} = \frac{H_{ec}}{\lambda}\tanh\left(\frac{a}{\lambda}\right) . \tag{7.2.2}$$

If $a \ll \lambda$, this approximates to

$$J_{tc} = \frac{H_{ec}}{\lambda}\frac{a}{\lambda} . \tag{7.2.3}$$

It is now assumed that, at the critical field strengths in the bulk super-conductor and in the thin plate, the surface current densities are the same. Hence, $J_c = J_{tc}$ and

$$\frac{H_{ec}}{H_c} \approx \frac{\lambda}{a} . \tag{7.2.4}$$

We have considered only large values of λ/a. To cover the entire range of plate thicknesses we write

$$\frac{H_{ec}}{H_c} = 1 + \frac{\lambda}{a} . \tag{7.2.5}$$

The basic relevance of these formulae for the enhancement of the critical field strength of thin superconductors has been proved by many experimenters. The results of some very early experiments on lead filaments are shown in Figure 7.2.1 (Pontius, 1937).

However, experiments on very thin filaments of lead and other soft superconductors indicate that the critical field eventually rises more rapidly with decreasing filament size than the simple theory suggests (Bean, 1962). This is due to the influence of the coherence length, a concept introduced in Section 3 and expanded later.

The very fine division of soft superconductors into thin films or wires will give considerably enhanced critical fields. If the synthesis of fine wires were the only way in which high-field superconductivity could be achieved, this would probably already have been achieved commercially. However, nature produces finely divided superconductors with only nominal human aid; she does so by invoking what is called surface energy.

The concept of surface energy must first be invoked, however, to explain why the Meissner effect occurs and, of course, the Meissner effect is the very antithesis of a finely divided superconductor.

We have seen that the presence of a magnetic field with an associated negative magnetisation (flux exclusion) raises the energy of a superconductor: the presence of the critical field just raises the energy to that of the normal phase.

Consider the penetration of magnetic flux into a thin superconductor such as is shown in Figure 7.1.1. The energy of magnetisation per unit volume of the thin superconductor is clearly not as great as that of a bulk superconductor for a given field. Thus a superconductor in a magnetic field in trying to reach a state of lowest energy should tend to divide into a laminated structure consisting of thin superconducting regions separated by yet thinner normal regions, for such a configuration would have a smaller net magnetisation and consequently a lower energy than an undivided superconductor. The expulsion of all magnetic flux from the bulk of a pure soft superconductor in the Meissner effect demonstrates clearly that this does not happen. This implies that a positive energy must be required to form the superconducting–normal interface in superconductors exhibiting the Meissner effect. How such a surface energy arises we shall now see.

In Section 3.1 the concept of a coherence length was introduced, this being the typical distance over which electrons can pair and thus become superconducting. In pure soft superconductors it transpires that the coherence length is many times greater than the penetration depth. At the surface, or any superconducting–normal interface, of a superconductor, electrons find themselves in a particularly unfavourable

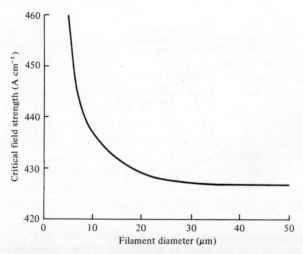

Figure 7.2.1. The critical field strength of lead filaments as a function of their thickness.

position for pairing, for there are no potential partners available on the normal side of the boundary. It is clear that this results in a finite rate of increase of superconducting electron pairs in the depth of the super-conductor. The typical distance at which the superconducting electron density (order parameter) has been built up to its value in the bulk of the superconductor is the coherence length ξ.

The finite rate of increase of the order parameter is in marked contrast with the assumption of Section 5.2. There, the penetration depth was calculated by assuming that a superconducting electron density corresponding roughly to the number of valence electrons per atom existed right out to the boundary of the superconductor. This would evidently account in a large part for the rather low value calculated for the penetration depth of lead, compared with the experimental value.

There is another and more important consequence of the finite variation of the density of superconducting electron pairs at a superconducting–normal boundary. Consider Figure 7.2.2 which shows the variation of the order parameter and the penetration of a magnetic field into a super-conductor. For the simplicity of a qualitative argument, the curves representing field penetration and variation of order parameter are reduced to the dashed lines. These divisions correspond to the coherence length ξ and the penetration depth λ. They divide the superconductor into three regions.

In region A the order parameter is at its maximum value. If this situation prevailed out to the surface, the energy of the whole super-conductor would be that of the superconducting phase. However, to a depth ξ below the surface the material is effectively not superconducting so that, per unit surface area, the energy is *increased* by $\frac{1}{2}\mu_0 \xi H_c^2$.

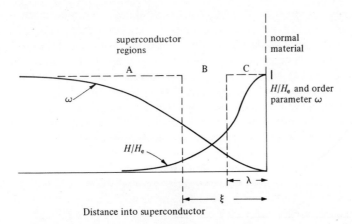

Figure 7.2.2. The qualitative spatial variation of the order parameter (density of superconducting electron pairs) and of magnetic flux as functions of depth below the surface of a superconductor.

In region C an external field penetrates to a depth λ. Here then the magnetisation of the superconductor is less than it would be if there were no penetration at all. This *reduces* the energy of magnetisation per unit surface area by $\frac{1}{2}\mu_0\lambda H_e^2$.

Thus together these two effects account for an increase in energy per unit area of

$$\tfrac{1}{2}\mu_0(\xi H_c^2 - \lambda H_e^2). \tag{7.2.6}$$

If $\xi > \lambda$, there will be an increase in the energy of the superconductor upon the formation of a superconducting–normal surface for all values of the external field up to H_c.

This surface energy is pivotal in the behaviour of superconductors. If it is positive, energy is required to form a superconducting–normal interface, and there being no source of energy a Meissner effect is possible. If there is zero or negative surface energy, a magnetised superconductor would tend to break up into alternate thin superconducting and infinitesimal normal regions. Because of the more complete penetration of magnetic flux into the thin superconducting regions the energy of magnetisation would be lower, a state to which the superconductor as a whole would tend. In superconductors displaying the Meissner effect this does not occur, showing that a positive surface energy exists.

The association of mechanical softness with the Meissner effect mentioned in Section 5.1 is now understandable. The Meissner effect arises through a positive surface energy. This is the result of a long coherence length ξ, for the formation of superconducting electron pairs. Because the pairing of electrons is accomplished via a phonon (sound wave) interaction, the regularity of the lattice has a strong influence on the coherence length. Soft metals are most easily obtained in a pure strain-free state with a regular lattice, which is essential for a long coherence length. On the other hand alloys, compounds, and impure strained metals will have in general a less regular lattice and therefore shorter coherence length. These materials have the characteristic of remaining superconducting in high magnetic fields and showing only a weak reversible Meissner effect. It is this property, together with a high critical temperature and high overall current density, which makes some superconductors ideally suited for use in high-field devices. How high an overall current density can be achieved in these materials will be seen in a subsequent section.

7.3 Superconductors of the first and second kinds†

Superconductors having a positive surface energy and exhibiting a Meissner effect at fields up to the bulk critical field H_c are termed type I super-conductors. As we have just seen they are usually the soft pure metals: they have a coherence length greater than the penetration depth.

† Goodman (1964).

Type II superconductors by contrast have short coherence length and a negative surface energy. They exhibit the Meissner effect only in a range of field strengths not extending up to the bulk critical field. Type II superconductors, because of their tendency to form a structure of mixed normal and superconducting regions of small dimension, remain super-conducting in high fields, above their bulk critical field. Their ability to retain superconductivity in high fields makes them essential to the construction of magnets and all high-field superconductive devices. Because of their importance we shall consider type II superconductors in some detail. Firstly, we shall develop a simple theory of their magnetic behaviour.

We shall suppose that a type II superconductor has formed within its bulk a laminated structure of alternating normal and superconducting layers.

In order to predict the gross magnetic behaviour of such a structure it is necessary to determine the equilibrium thickness of the normal and superconducting laminae. To do this we calculate, in general terms, the energy of the structure and minimise it with respect to the thickness of the superconducting laminae (see Figure 7.3.1).

The magnetic energy of this structure consists of the following components.

1 The basic potential energy of the superconducting laminae which is $-\frac{1}{2}\mu_0 H_c^2$ per unit volume referred to the energy of the normal state.

2 The basic energy density of the normal state, here assumed to be zero.

3 The potential energy of the superconducting state arising from its magnetisation. If there were no flux penetration this potential energy density would be $\frac{1}{2}\mu_0 H_e^2$ per unit volume.

4 The reduction in potential energy of magnetisation due to the penetration of the flux into the superconducting layer. In Section 5.1 this small reduction was neglected, because, for bulk superconductors

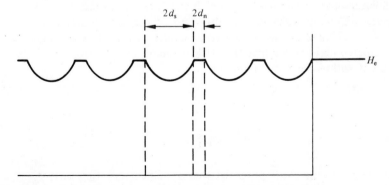

Figure 7.3.1. The assumed variation of magnetic flux density (field strength) within a laminated superconductor (see text).

of large volume, the penetration of the field to a depth of about 10^{-5} cm is of no consequence. Now, however, we are considering laminae whose thickness is of this order and we must therefore calculate the reduction in the energy of magnetisation caused by the penetration. Why the potential energy of the system is reduced by such penetration should be mentioned. For, it might at first sight seem that the acceleration of electrons brought about by the flux penetration had increased the energy of the superconductor. So indeed it has. But the potential energy of magnetisation has been reduced and it is the potential energies of the structure whose equilibrium we seek. The essence of the calculation of the equilibrium energy of the system is the determination of the effective potential energy of magnetisation of the laminated structure including the following.

5 The fifth and final component of energy arises from the finite rate of increase of correlated electron pairs from the surface towards the interior of each superconducting lamina. This in effect amounts to a decrease in the width of the superconducting lamina by an amount of 2ξ. This therefore represents an increase in energy of $\frac{1}{2}\mu_0\xi H_c^2$ per unit area of a lamina. This is the source of the surface energy of the superconductor. It will be noticed that our calculation is developing in much the same way as did our simple criterion above for the surface energy, which we found to be given by $\frac{1}{2}\mu_0(\xi H_c^2 - \lambda H_e^2)$.

The difference now is that we are going to calculate the energy of magnetisation more exactly. Apparently, however, we are not going to calculate the effect of the finite increase in density of correlated pairs more exactly: we have grouped the effect into a factor $\frac{1}{2}\mu_0\xi H_c^2$. The reason for doing this is the complexity of the spatial variation of the correlation function. The justification for doing so is that we shall be particularly interested in very short coherence lengths, smaller for the most part than the thickness of the superconducting laminae.

Our first task is to calculate the exact potential energy of magnetisation of the superconducting laminae into which the field has penetrated. We shall use the following notation: $2a$, thickness of the superconducting laminae, $2b$, thickness of the normal laminae, $H(x)$, local field strength within a superconducting laminae, H_e, field strength external to the whole structure and within the normal laminae,

$$h = \frac{H_e}{H_c}, \qquad\qquad p = \frac{a}{\lambda},$$

$$q = \frac{\xi}{\lambda}, \qquad\qquad \text{and} \quad \frac{a}{a+b} = \chi.$$

We can calculate the exact potential energy of magnetisation in two ways. We may calculate the kinetic and magnetic energy of the electrons $(\frac{1}{2}LI^2 + \frac{1}{2}nmv^2)$ and subtract it from the energy of magnetisation which would obtain if there were no flux penetration (London, 1950, p.125 ff.). Alternatively we may calculate immediately the energy of magnetisation by integration of the local energy of magnetisation across the laminae. We shall use the latter method. This in fact has been done in Equation (7.1.4), which gives as the potential energy of magnetisation per unit volume

$$\epsilon_m = \tfrac{1}{2}\mu_0 H_e^2 \left[1 - \frac{\lambda}{a}\tanh\left(\frac{a}{\lambda}\right)\right] . \tag{7.3.1}$$

Hence, per unit surface area of the superconducting lamina

$$\epsilon_m = \tfrac{1}{2}\mu_0 H_e^2 2a \left[1 - \frac{\lambda}{a}\tanh\left(\frac{a}{\lambda}\right)\right] . \tag{7.3.2}$$

We can now write down the potential or free energy of the structure of superconducting and normal laminae. Taking unit surface area and unit thickness of the structure, the free energy is

$$G(H_e) = -\tfrac{1}{2}\mu_0 H_c^2 \frac{a}{a+b} + G_n \frac{b}{a+b}$$
$$+ \tfrac{1}{2}\mu_0 H_e^2 \frac{2a}{2a+2b}\left[1 - \frac{\lambda}{a}\tanh\left(\frac{a}{\lambda}\right)\right] + \tfrac{1}{2}\mu_0 \xi H_c^2 \frac{1}{a+b} . \tag{7.3.3}$$

These are respectively the five terms enumerated above. The third and fourth terms are grouped together in the brackets. G_n is assumed to be zero but is included in the second term for completeness.

This expression reduces to

$$G(H_e) = G_n(1-\chi) + \tfrac{1}{2}\mu_0 H_c^2 \chi\left[-1 + h^2\left(1 - \frac{\tanh p}{p}\right) + \frac{q}{p}\right] . \tag{7.3.4}$$

Now the arrangement of the laminae will tend to be such that the potential energy of the system will be as low as possible. However many superconducting lamina there may be, the energy $G(H_e)$ will be lowest if $\chi = 1$, that is if the normal laminae are of infinitesimal thickness. In fact we shall see later that the normal laminae cannot have a thickness less than the coherence length ξ, but for our simple theory we shall set $\chi = 1$.

Then,

$$G(H_e) = \tfrac{1}{2}\mu_0 H_c^2\left[-1 + h^2\left(1 - \frac{\tanh p}{p}\right) + \frac{q}{p}\right]. \tag{7.3.5}$$

This free energy will tend to minimise itself by changing the thickness of the superconducting laminae according to the magnitude of the applied field. To find the equilibrium condition for p, we differentiate the free energy with respect to p and equate the derivative to zero. This gives, for the equilibrium condition,

$$\frac{q}{h^2} = \tanh p - p\,\mathrm{sech}^2 p . \tag{7.3.6}$$

We can examine this expression for the value of q which distinguishes a type I from a type II superconductor. A type I superconductor exhibits the Meissner effect in fields up to the bulk critical field, that is for $h = 1$. The complete expulsion of field is identified in Equation (7.3.6) with a very large value of p. In that case,

$$\frac{q}{h} = 1 \ , \tag{7.3.7}$$

which means that, if $h = 1$, then $q = 1$ also. Thus, if $q \geqslant 1$, the superconductor will be type I. If $q < 1$, the superconductor will be of type II. It will display a Meissner effect only up to a field strength given by

$$h = q^{\frac{1}{2}} \ ,$$

that is

$$H_e = H_c \left(\frac{\xi}{\lambda}\right)^{\frac{1}{2}} = H_{c1} \ . \tag{7.3.8}$$

H_{c1} is called the lower critical field of the type II superconductor.

This result could also have been inferred directly from Equation (7.2.6). What follows could not have been so inferred, however.

If $q < 1$ and $H_e > H_{c1}$, solutions to Equation (7.3.6) exist for finite values of p. As h is increased p decreases, as more but thinner laminae are squeezed into the structure. A limit to the density of laminae is reached when the energy per unit volume of the structure is equal to that of the normal state. This condition is expressed by

$$\tfrac{1}{2}\mu_0 H_c^2 \left[-1 + h^2 \left(1 - \frac{\tanh p}{p} \right) + \frac{q}{p} \right] = 0 \ . \tag{7.3.9}$$

Combining this with Equation (7.3.6) to eliminate q we get

$$p = \tanh^{-1}\left(\frac{1}{h''}\right) \ , \tag{7.3.10}$$

where $h'' = H_{c2}/H_c$, H_{c2} being the upper critical field of the superconductor.

If we insert this value for p into Equation (7.3.6), the equation defining H_{c2} becomes

$$q = \frac{H_{c2}}{H_c} + \left[1 - \left(\frac{H_{c2}}{H_c}\right)^2 \right] \tanh^{-1}\left(\frac{H_c}{H_{c2}}\right) = \frac{\xi}{\lambda} \ . \tag{7.3.11}$$

For very small values of q this expression can be satisfactorily approximated by

$$\frac{H_{c2}}{H_c} \approx \frac{2}{3}\left(\frac{\lambda}{\xi}\right) \tag{7.3.12}$$

and by combining this with Equation (7.3.8) we get

$$\frac{H_{c2}}{H_{c1}} \approx \frac{2}{3}\left(\frac{\lambda}{\xi}\right)^{\frac{1}{2}} \ . \tag{7.3.13}$$

The magnetic behaviour of a superconductor is most easily characterised by its magnetisation. We already know that a type I superconductor is completely diamagnetic up to its bulk critical field H_c. A type II superconductor is completely diamagnetic only up to its lower critical field above which field its magnetisation rapidly diminishes.

Between H_{c1} and H_{c2}, the magnetisation is given by Equation (7.1.3), that is

$$\frac{M}{H_c} = -h\left(1 - \frac{\tanh p}{p}\right). \tag{7.3.14}$$

Corresponding to various values of h, p is found from Equation (7.3.6), for a given value of the surface energy parameter q. The resulting magnetisation curves are shown in Figure 7.3.2a.

It will be noticed that as q decreases the magnetisation curves become shallower and extend to higher fields. Also it will be realised that the bulk critical field strength of a type II superconductor is not a directly measurable quantity. It is still a meaningful parameter, however, for it is a measure of the energy difference between the unmagnetised superconducting and normal phases of the superconductor. During the increase of the external field from zero to H_{c2} the energy of the superconductor is raised by an amount

$$G_n - G_s = \tfrac{1}{2}\mu_0 H_c^2 .$$

This must be equivalent, however, to the energy of magnetisation over the same field interval; thus

$$\mu_0 \int_0^{H_{c2}} M \, dH = \tfrac{1}{2}\mu_0 H_c^2 .$$

The area under the magnetisation curve is therefore identical with $\tfrac{1}{2}\mu_0 H_c^2$, the energy difference between the phases.

The magnetisation curves of Figure 7.3.2a are called the intrinsic characteristic of the type II superconductor to distinguish it from other characteristics described later. The intrinsic magnetisation is completely reversible, as indeed it must be in order that a thermodynamic treatment be applicable.

Having thus constructed a fairly simple model of a quite complex physical phenomenon we must now regrettably examine its shortcomings and the reasons for its disagreement with what is actually measured. Typical real magnetisation curves of a type II superconductor are shown in Figure 7.3.2b. The most marked discrepancy between our model and actuality is the residual finite magnetisation at H_{c2} displayed by the laminar model but not observed in practice. The reason for this and other less important differences are as follows.

We assumed a laminar structure for ease of calculation. The principle of the minimisation of energy, however, leads, at least by qualitative arguments, to a different arrangement of normal and superconducting regions in the structure. Because the surface energy is negative in the type II superconductor, the energy of the system would be a minimum if the interface area between normal and superconducting regions were a maximum.

This would be obtained if the normal regions took the form of cylinders rather than laminae. Flux would then penetrate the structure in the form of cylindrical bundles, the radius of these being not less than the coherence length ξ. These flux bundles each would contain the minimum flux permissible, namely one quantum, and would be surrounded by circulating currents. (The reason that the flux would be a minimum is given in Section 12 in connection with the quantum magnetometer.)

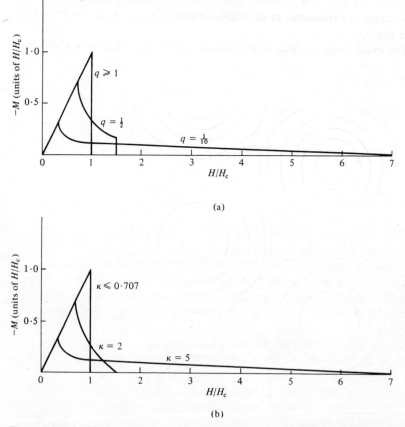

(a)

(b)

Figure 7.3.2 (a) The calculated magnetisation plotted as a function of field strength for type II superconductor with an assumed laminar structure. The independent parameter of the family of curves is the surface energy q. For $q > 1$ the curves represent the magnetisation of a type I or soft superconductor. (b) The variation of magnetisation as a function of field strength for type II superconductors.

These flux bundles are called fluxoids, fluxons, or vortices, the latter name arising from the circulating currents surrounding each flux quantum. These vortices would exert a repelling force on each other, because the currents of neighbouring vortices would flow in opposite directions at points midway between them. The vortices would then position themselves so that the forces were mutually, statically, and stably balanced. This would be achieved in a triangular array such as is shown in Figure 7.3.3 (Kleiner *et al.*, 1964).

The detailed and exact theory of the penetration of flux into a type II superconductor indeed predicts such a structure, consisting of normal cores of radius equal to the coherence length ξ, surrounded by a superconducting matrix into which the field penetrates with a characteristic length λ. Figure 7.3.4 shows the profiles of current density and flux in such a structure. The structure is known as the mixed state. Its importance is paramount in the exploitation of superconductivity in high-field devices.

The exact theory is due to Abrikosov (1957) who originally predicted a

Figure 7.3.3. The actual variation of field strength and order parameter in a type II superconductor. The triangular array achieves a lower total free energy than the simple laminar model, and is therefore the array into which the flux pattern will settle in a type II superconductor. The numbers refer to the order parameters.

square array of vortices. The triangular array shown in Figure 7.3.3 was calculated by Roth and colleagues (Kleiner *et al.*, 1964). Abrikosov's theory was based on a precise two-fluid model of phenomenological super-conductivity by Ginsburg and Landau (1950; see also Harden and Arp, 1963). In their theory the order parameter is treated as a complex quantity to obtain spatial and time dependence. They found that the properties of a superconductor could be characterised by a constant

$$\kappa = \frac{2(2)^{1/2} e H_c \lambda^2}{2\pi h} , \qquad (7.3.15)$$

where h is Planck's constant.

Type II superconductivity is characterised by $\kappa > 1/(2)^{1/2}$. [We can identify this with our criterion for type II superconductivity that $\lambda/\xi > 1$. But, because we chose a simple laminar model, we find that there is an error of a factor of $(2)^{1/2}$.]

Thus it is found that $\kappa = \lambda/\xi$ and for type II superconductivity

$$\frac{\lambda}{\xi} > \frac{1}{(2)^{1/2}} .$$

Stated in approximate form, Abrikosov's expressions for H_{c1} and H_{c2} are

$$H_{c2} = (2)^{1/2} \kappa H_c , \qquad (7.3.16)$$

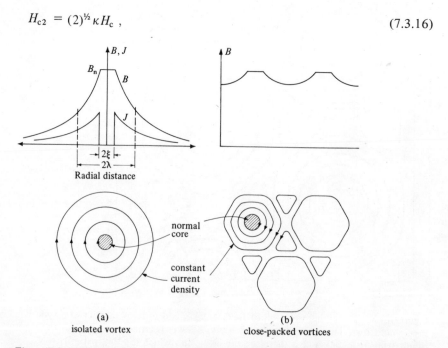

(a)
isolated vortex

(b)
close-packed vortices

Figure 7.3.4. The profiles of current density and flux density through one vortex of the triangular flux array of a type II superconductor.

$$H_{c1} = \frac{H_c}{\kappa}\left(\ln\kappa + 0\cdot08\right) .$$

(7.3.17)

The coherence length ξ and penetration depth λ are dependent on the electronic properties of the normal state of a superconductor. Thus the coherence length is related closely to the electronic mean free path. In the limit of a very short mean free path the coherence length is equal to it. Similarly the penetration depth is a function of the number of free electrons per unit volume, which is a characteristic of the normal state. Because of these dependences it is possible to define the constant κ and hence the upper critical field in terms of the properties of the normal state. An expression for κ was derived by Gorkov. This was developed by Shapoval (1962) into an expression for the upper critical field H_{c2}. This is given by

$$H_{c2} = \frac{3}{2\pi}\frac{e\gamma T_c}{k\sigma} ,$$

(7.3.18)

where γ is the electronic specific heat, σ the normal conductivity at low temperature, and k is Boltzmann's constant. The collective theories of Ginsburg, Landau, Abrikosov, and Gorkov are the firm basis of the description of type II superconductors. Together the theories carry the acronym GLAG theory (Lynton, 1964).

The direct experimental observation of the mixed state was not achieved until 1966 (Trauble and Essmann, 1968). The technique needed to observe effects on almost an atomic scale at low temperatures is basically simple but refined. The dimensions involved are about 10^{-6} to 10^{-7} cm for this is the typical value of the coherence length of type II materials. The experiment is performed thus. A sample of type II material is cooled to below its critical temperature and a magnetic field is applied in such an orientation that it penetrates uniformly into the sample. The space surrounding the sample is evacuated and iron is evaporated from a source above the sample. Condensed iron vapour is drawn along the field lines towards the surface of the sample where it rests clustered around the points of exit of the fluxoids. The field is slowly reduced to zero and the sample allowed to warm up. When warm a layer of carbon is evaporated over the surface after which the vacuum is released. Varnish is next sprayed onto the carbon which can then be lifted from the sample with the clusters of the iron particles and viewed in an electron microscope. The pattern seen is shown in Plate I. The triangular array of fluxoids is clearly visible. Also, a number of dislocations can be seen in the fluxoid lattice. These dislocations may be intrinsic to the fluxoid lattice itself or they may be due to inhomogeneity in the lattice of the sample. The latter are of great importance in the ability of a type II superconductor to carry a bulk transport current.

7.4 Pauli paramagnetism†

The normal conduction electrons of a metal have the ability to become polarised by a magnetic field. This effect, arising from the property of spin possessed by the electrons, is known as paramagnetism. The superconducting electron pairs do not possess a paramagnetic moment, however. The electrons of a bound pair have equal but opposite spin and the net moment in a magnet field is zero. The effect of these properties is that the free energy of the normal phase of a superconductor is reduced by its paramagnetism whilst the free energy of the superconducting phase is unaltered by it.

The paramagnetic susceptibility of most metals is extremely small but in some type II superconductors at very high fields it can become comparable in magnitude with the diamagnetism of the superconducting mixed state. In these cases the upper critical field H_{c2} is found experimentally to be less than would be expected to follow from other characteristics of the material. For instance, in Figure 7.4.1 is shown the experimental curve of upper critical field as a function of temperature for the intermetallic compound vanadium–gallium (V_3Ga): it is shown as the solid curve (Ginsburg and Landau, 1950, Harden and Arp, 1963).

Since $H_{c2} = (2)^{1/2} \kappa H_c$, the temperature variation of H_{c2} will follow that of H_c and k. From Section 5.1

$$H_c = H_0 \left[1 - \left(\frac{T}{T_c} \right)^2 \right] . \tag{7.4.1}$$

Assuming for the moment that κ is independent of temperature, we can write

$$H_{c2}(T) = H_{c2}(0) \left[1 - \left(\frac{T}{T_c} \right)^2 \right] , \tag{7.4.2}$$

where $H_{c2}(0)$ is the upper critical field at absolute zero. The slope of the (H_{c2}, T) curve at T_c is then

$$\frac{dH_c}{dT} = -2H_{c2}(0) T_c^{-1} \tag{7.4.3}$$

and it follows that

$$H_{c2}(T) = -\tfrac{1}{2} T_c \left(\frac{dH_{c2}}{dT} \right)_{T=T_c} \left[1 - \left(\frac{T}{T_c} \right)^2 \right] . \tag{7.4.4}$$

Because of a small temperature variation of κ, $H_{c2}(0)$ is more accurately written

$$H_{c2}(0) = -0\cdot69 \left(\frac{dH_{c2}}{dT} \right)_{T=T_c} , \tag{7.4.5}$$

† Montgomery and Wizgall (1966); Hake (1965); Clogston (1962).

but this is not important in the present discussion. Thus with a knowledge of only T_c and $dH_{c2}/dT_{T=T_c}$ the dashed curve of Figure 7.4.1 can be drawn. The theoretical value of the upper critical field in the absence of paramagnetic limitation (designated H_{c2}^*) is evidently greater than the actual value. The difference is greatest at low temperatures.

An approximate calculation of the effect of paramagnetic limitation of the upper critical field can be made as follows.

The free energy of the normal state is decreased by its paramagnetism. This decrease is given by

$$\Delta G_{np} = \tfrac{1}{2}\chi_p\mu_0 H^2 , \qquad (7.4.6)$$

where χ_p is the paramagnetic susceptibility. The magnetisation of a type II superconductor near its upper critical field in the absence of paramagnetism is given approximately by

$$M = -(H_{c2}^* - H)[1 \cdot 18(2\kappa^2 - 1)]^{-1} , \qquad (7.4.7)$$

where H_{c2}^* is the upper critical field in the absence of paramagnetism, H is a given field $< H_{c2}^*$, and $\kappa \gg 1$ and is as defined before. Consider Figure 7.4.2 showing the magnetisation curve of a type II superconductor

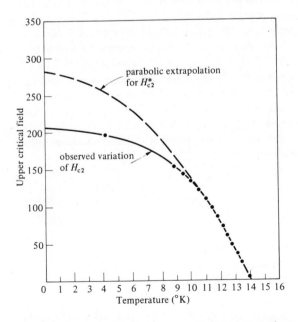

Figure 7.4.1. The variation of the upper critical field H_{c2} of compound V_3Ga as a function of absolute temperature. The solid curve is as experimentally determined: the dashed curve represents H_{c2}^* and is the parabolic extrapolation of the curve from T_c. The difference between the curves represents the reduction in H_{c2}^* caused by Pauli paramagnetism.

in the absence of any paramagnetic limitation. At H the energy of the superconductor will be

$$G_s = G_n - a \,, \tag{7.4.8}$$

where a is the shaded area under the magnetisation curve. This area is given by

$$a = \int_{H_{c2}^*}^{H} \mu_0 M \, dH \,. \tag{7.4.9}$$

Substituting for M from Equation (7.4.7), we obtain

$$a = -\tfrac{1}{2}\mu_0 (H_{c2}^* - H)^2 [1 \cdot 18(2\kappa^2 - 1)]^{-1} \,. \tag{7.4.10}$$

If $a = \Delta G_{np}$, the energies of the superconducting and normal states will be equal. Since χ_p is a small quantity, H_{c2} is appreciably smaller than H_{c2}^* only if κ is large.

Equating expressions (7.4.6) and (7.4.10) we get a value of H which is the upper critical field in the presence of paramagnetism:

$$H_{c2} = H_{c2}^* \{1 + [1 \cdot 18\chi_p (2\kappa^2 - 1)]^{\frac{1}{2}}\}^{-1} \,. \tag{7.4.11}$$

Typical values of χ_p are in the range $10^{-4} - 10^{-5}$. The reduction in H_{c2} will therefore only be significant if κ is large.

Pauli paramagnetism imposes an ultimate limit to the critical field that could be expected of a superconductor. For instance, suppose a superconductor is so finely divided that, according to the principles discussed in Section 7.1 and if we ignore the effect of ξ in limiting the minimum dimension, its critical field in the absence of paramagnetism would be infinite. Owing to paramagnetism, an applied field decreases the energy of the normal state. This decrease will, at a sufficiently high field, completely compensate the difference in free energies between superconducting and

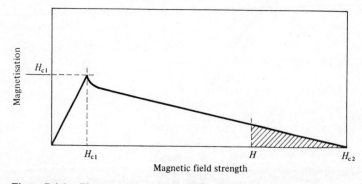

Figure 7.4.2. The intrinsic magnetisation of a type II superconductor as a function of field strength. The shaded area represents the energy difference between the superconducting phase at H and the normal phase.

normal states. Then,

$$\tfrac{1}{2}\mu_0 H^2 \chi_p = G_n - G_s .$$ (7.4.12)

Both χ_p and $G_n - G_s$ are related to the density of energy states near the Fermi energy and thereby to the critical temperature of the superconductor. It transpires that the limiting critical field imposed by paramagnetism is given by

$$H_p \approx 14920 T_c ,$$

in which H_p will be in amperes per centimetre and T_c in degrees Kelvin (Clogston, 1962).

There are other effects which reduce the effect of Pauli paramagnetism. In their absence, however, it would be expected that a superconductor having a critical temperature of 20°K would not have an upper critical field exceeding about 300 kA cm^{-1}.

7.5 Surface superconductivity†
The conditions at the interface between a type II superconductor and an insulator or normal metal lead to surface superconductivity at fields above H_{c2}. The last trace of surface superconductivity disappears at a field $H_{c3} = 1 \cdot 69 H_{c2}$.

The surface current flowing in this range of fields is small compared with the Silsbee surface current at H_{c1} (see Section 2.1). The magnitude of the current depends on the ratio ρ_n / ρ_s, where ρ_n is the resistivity of the normal material and where ρ_s is the resistivity of the superconductor in the normal state. This ratio must be high for appreciable surface current: surface superconductivity above H_{c2} is normally only found in polished samples.

† Saint-James and de Gennes (1963).

Hard superconductors

8.1 Flux flow and flux pinning†

It may at first seem inevitable that a type II superconductor, having a high upper critical field H_{c2}, should also have a high current-carrying capacity. In fact this is not necessarily so. A suitably prepared type II super-conductor is unable to support a bulk transport current in the presence of a field greater than H_{c1}, oriented at right angles to the current.

To understand this it is necessary to refer to Figure 8.1.1a. This shows a number of current vortices surrounding cores of flux within a type II superconductor in a uniform external field. Clearly no net current flows along the superconductor in this case and the spatial density of vortices is uniform. By contrast in Figure 8.1.1b a transport current is flowing, producing asymmetry in the external field strengths on either side of the superconductor and a density gradient in the vortices. There is evidently a Lorentz force on each vortex caused by the interaction of the magnetic flux with the orthogonal transport current. (Alternately, the force can be considered as arising from the density gradient of the vortices.) This force

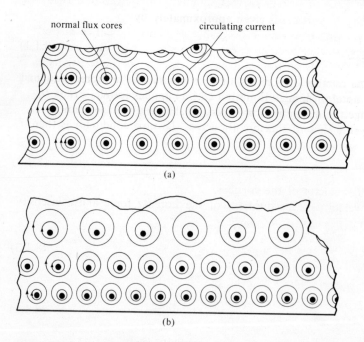

Figure 8.1.1. The spatial distribution of vortices in a type II superconductor under conditions (a) of uniform field strength and (b) of a gradient of field strength.

† Anderson and Kim (1964); Goodman (1964); Heaton and Rose-Innes (1964).

causes the vortices to flow across the superconductor from a region of high field to one of low field. The velocity at which the vortices traverse the superconductor is determined by the balance of the Lorentz force $J \times B$ and the viscous drag on the vortices arising from their motion through the superconductor. On reaching the low-field side of the superconductor the vortex currents decay and the flux quanta contained by the vortices are annihilated. Vortices are formed on the high side of the superconductor to make good this loss, which must be supported by an energy input from an external source.

The net power dissipation is proportional to the rate of annihilation of vortices. This manifests itself as a resistance in the superconductor, called the flow resistivity. A simple calculation yields a value for flow resistivity as a function of field strength and normal-state resistivity of the superconductor.

For a given transport current density J_t the rate at which vortices cross the superconductor is proportional to the density of the vortices. For a given applied field $H_e \gg H_{c1}$, the flux density within the superconductor is given closely by

$$B_i = \mu_0 H_e \,,$$

and the density of vortices is given approximately by

$$\sigma = \frac{H_e}{H_{c2}} \frac{1}{\pi \xi^2} \,, \tag{8.1.1}$$

where σ is the number of vortices, or flux quanta, per unit area; H_{c2} and ξ have their usual meaning.. Equation (8.1.1) applies for $H_{c2} - H \ll H_{c2}$, in which range the magnetisation is very small. The voltage gradient generated by the flow of flux quanta across the superconductor is then given by

$$E = \sigma \phi_0 v \tag{8.1.2}$$

where v is the velocity of the vortices.

Now the velocity is the quotient of the Lorentz force and a viscosity constant; thus,

$$v = \frac{J_t \phi_0}{\eta} \,, \tag{8.1.3}$$

whence the voltage gradient is given by

$$E = \frac{\sigma J_t \phi_0^2}{\eta} \,. \tag{8.1.4}$$

For a given transport current the flow resistivity is

$$\rho_f = \frac{E}{J_t} \,.$$

Thus,

$$\rho_f = \frac{\phi_0^2 \sigma}{\eta} = \frac{H_e}{H_{c2}} \frac{\phi_0^2}{\pi \xi^2 \eta} \; . \tag{8.1.5}$$

When $H_e = H_{c2}$ full normal resistivity is restored so that Equation (8.1.5) reduces to

$$\rho_f = \frac{H_e}{H_{c2}} \rho_n \; , \tag{8.1.6}$$

where ρ_n is the normal-state resistivity. If superconductors with strong Pauli paramagnetism are considered the denominator of Equation (8.1.6) is H_{c2}^*.

This dissipative behaviour is a property of type II superconductors and is most pronounced in annealed defect-free materials. *It is the presence of lattice irregularities in type II superconductors that 'pins' the vortices against movement under the influence of Lorentz forces and enables a dissipationless current to flow perpendicular to the magnetic flux vector.* Type II superconductors in which flux movement is inhibited by 'pinning' are called hard superconductors.

Although the picture of the movement of discrete bundles of flux is easy to appreciate at first sight, it is one that tends to become quickly open to question. For instance, the appearance of a voltage in a type II superconductor could be interpreted very simply as being caused by the development of resistive sections in the superconductor. However, the results of two sets of experiments prove that flux flow does indeed occur in type II superconductors.

In the first of these experiments the entropy carried across a superconductor by the vortices is measured (Otter and Solomon, 1966).

At the centre of a vortex in a type II superconductor lies a cylindrical normal region of radius roughly equal to the coherence length. The electrons in this normal region have all the characteristics associated with the normal state of the superconductor, including a higher entropy than those in the superconducting state. The formation of a vortex in the surface of a superconductor requires a quantity of heat per unit length of the vortex roughly given by

$$\Delta Q = \tfrac{1}{2} \pi \xi^2 \mu_0 H_c^2 \; . \tag{8.1.7}$$

For a type II superconductor having a coherence distance of about $2 \cdot 5 \times 10^{-6}$ cm and a bulk critical field of 800 A cm^{-1} the energy ΔQ is approximately 10^{-13} J cm^{-1}.

This heat is in fact the latent heat of magnetisation mentioned previously in Section 5.1.

When a vortex is annihilated, as for instance on arriving at the low-field side of a superconductor, this latent heat is released. The experiment consisted of the measurement of this heat.

A sample of an alloy of indium with 40 at.% of lead was secured between two aluminium rods in vacuum. The sample was 4 cm long by $\frac{1}{2}$ cm wide and 0·16 cm thick. The aluminium rods served as current leads and thermal anchors, their lower ends being immersed in liquid helium. The configuration of the sample was as shown in Figure 8.1.2. The voltages along each edge of the sample were measured by small potential probes and the temperatures at each side of the centre of the sample were measured by resistance thermometers. Flux flow was produced by the interaction of a field oriented at right angles to the face of the sample and a transport current through it. The data of the experiment are summarised in two sets of curves. Consider first Figure 8.1.3. In this figure is plotted the temperature difference across the width of the sample as a function of the voltage gradient along it, divided by the thermal conductivity. The curves were obtained by varying the transport current in the range 1–10 A.

Suppose n vortices, each carrying a quantity of heat ΔQ, leave unit length of the sample per second. The heat transported $n \Delta Q t$ must flow back across the sample under steady-state conditions, driven by the thermal gradient $\Delta T_r / w$.

Thus,

$$n \Delta Q t = \frac{K \Delta T_r t}{w} , \qquad (8.1.8)$$

where t is the thickness, w the width of the sample, and K is the thermal conductivity. n is known from the voltage gradient, however, because

$$n \phi_0 = E , \qquad (8.1.9)$$

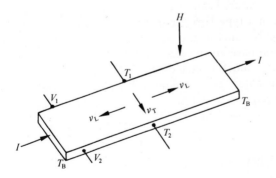

Figure 8.1.2. Configuration of the sample used to demonstrate the entropy transport of flux flow in a type II superconductor. The ends of the sample are held at a base temperature T_b. Vortex flow takes place transversely (v_T) because of Lorentz force, and longitudinally (v_L) because of temperature gradient. The transverse voltage gradient is $V_1 - V_2$; the transverse temperature gradient is $T_1 - T_2$ and the longitudinal temperature gradient is $\frac{1}{2}(T_1 + T_2) - T_B$.

Therefore

$$\Delta T_r = \frac{\Delta Q w E}{\phi_0 K} .$$

(8.1.10)

The slope of the curves in Figure 8.1.3 gives $\Delta T_r K/E$ at $4 \cdot 2°$K and 800 A cm^{-1}. Hence,

$$\frac{\phi_0}{\Delta Q w} = 2 \cdot 5 \times 10^{-3} \text{ mV degK W}^{-1} \text{ cm}^{-1}$$

and

$$\Delta Q = 1 \cdot 6 \times 10^{-13} \text{ J cm}^{-1}.$$

This is in agreement with the value calculated from the coherence length and the bulk critical field.

At $2 \cdot 6°$K it is seen that $\Delta Q w/\phi_0$ is considerably greater. This increased heat transport comes from the increased bulk critical field at the lower temperature.

These results show that vortices carry entropy across a type II superconductor and are driven by the Lorentz force $J \times B$. An extension of the same experiment shows that vortices can also be driven by a thermal gradient.

The transverse voltage difference across the width of the sample at one end close to an aluminium support was measured and plotted against the transport current in the sample. This plot is shown in Figure 8.1.4 for

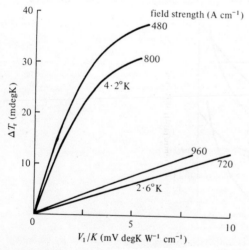

Figure 8.1.3. Transverse temperature difference across a sample of type II superconductor as a function of the longitudinal voltage per unit thermal conductivity. The temperature difference is a measure of entropy transport by the vortices and the voltage is a measure of their velocity. The curves show that these are mutually proportional and lead to a value for the heat transport associated with a flux quantum.

transport currents flowing in either direction. Two aspects of the curves deserve immediate comment: (1) no transverse voltage appears until a current of $3\frac{1}{2}$ A flows—this is due to some inherent flux pinning by unavoidable defects in the alloy, (2) at currents above $3\frac{1}{2}$ A, of either polarity, a transverse voltage appears which rises rapidly with current. These voltages each have one component which is dependent on the polarity of the current and one which is independent. The latter arises from the flow of vortices along the sample, driven by a temperature gradient. The temperature difference between the centre of the sample and the end where it is anchored to the aluminium post is also plotted. It is seen that the component of transverse voltage independent of the polarity of the current is proportional to this temperature difference. Hence the velocity of drift of the vortices is proportional to the temperature difference. From these observed drift velocities it was deduced that a temperature gradient of 1 degK cm^{-1} exerts the same force on the vortices as a current density of 3 A cm^{-2} in a flux density of 0·06 Wb m^{-2} at 4·2°K.

The second of the experiments confirming the existence of flux flow is the d.c. transformer. This is dealt with more fully in Section 11.1 but it suffices to say here that the d.c. transformer consists of two films of type II superconductor separated by a very thin insulating layer. If the vortices in the films are set into motion by a transport current in one film, a d.c. voltage will be generated by their movement in the other film.

The concept of flux flow tends to become less acceptable as greater

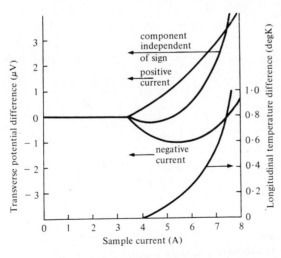

Figure 8.1.4. Transverse potential difference across a sample of type II superconductor and longitudinal temperature difference as functions of longitudinal sample current. The component of potential difference which is independent of polarity arises from the force on the vortices caused by the temperature gradient.

thought is devoted to it. At one period in the development of the concept of flux flow, the tenor and content of discussion resembled the 15th-century arguments about angels and pins. One of the more difficult aspects of flux flow concerns the processes occurring in a circuit consisting of a battery, a normal resistive conductor, and a type II superconductor. If flux flow does not occur, the battery voltage is used solely to overcome the resistance of the normal conductor. When flux flow commences, as for instance it will if a magnetic field is applied, vortices move continuously across the superconductor. The first question to come to mind is 'what becomes of the flux within the circuit?'. Quite clearly nothing becomes of it. It remains constant despite the movement of vortices across the superconductor. One may, for the convenience of a physical picture, imagine the vortices to be bundles of flux lines as shown in Figure 8.1.5. Then movement of the bundles does not result in movement of the main flux. The flux lines will continuously break and rejoin to allow the vortices to move. The movement of the vortices creates a back e.m.f. which reduces the net voltage available to drive current through the normal conductor.

The formation of a vortex can be imagined as a process by which a region of initially uniform field shrinks slightly, thereby producing a circumscribing current. This current then exerts an inwardly directed force on the flux bundle, which contracts until an equilibrium configuration having a minimum energy is obtained. The process is somewhat analogous to the 'pinch effect' in electrical discharges.

In order to prevent the movement of the vortices, sites must exist within the superconductor at which the vortices become 'pinned'.

How this flux pinning arises can be understood by considering the

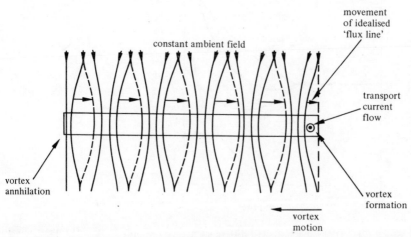

Figure 8.1.5. Visualisation of the movement of flux lines of a vortex. The flux lines are imagined to break and reform continuously on successive vortices so that, although the latter move to the left, the ambient magnetic flux does not move.

reduction in free energy of a vortex in the presence of a long cylindrical hole in the superconductor.. Let this vortex occupy one of two regions A or B in a superconductor. In the middle of region B there is a cylindrical hole of radius ξ through the superconductor. When in region A, the normal core of the vortex increases the free energy of the superconductor by an amount

$$\Delta\epsilon = \tfrac{1}{2}\pi\xi^2\mu_0 H_c^2 \text{ joules per unit length of flux core.} \qquad (8.1.11)$$

If the core is situated in the hole in region B, then in the presence of a vortex the free energy of region B is less by an amount $\Delta\epsilon$ than that of region A. This then is the energy difference favouring the location of the vortex in region B: it is the pinning energy of the hole in region B.

As we have already seen, Equation (8.1.11) leads to a typical value for $\Delta\epsilon$ of 10^{-13} J cm^{-1} per flux quantum. A cylindrical hole of radius ξ is an unlikely and extreme case of an imperfection in a lattice. There will, however, be regions in a strained or otherwise inhomogeneous state in a type II superconductor where the correlation energy is lowered appreciably. In these regions, the free energy of the superconductor will be higher than elsewhere and in an extreme case could be that of the normal material. The 'strength' of the pinning centre will depend on its free energy: the closer its free energy is to the normal-state free energy, the stronger will be the 'pinning' effect on a vortex core located in it.

Evidently, imperfections in a metal lattice may be extended regions, rather more like fissures than holes, they may run along grain boundaries, or they may be isolated regions of quite small size. Extended imperfections will tend to pin more than one flux quantum simultaneously. In these cases, the normal core region of a vortex may contain a large, but always integral, number of flux quanta. Isolated regions strung out along a flux line will give weaker pinning than our idealised cylindrical hole.

Although ensuing arguments consider single flux quanta, it is necessary to remember that 'bundles' containing one to thousands of flux quanta linked by one distribution of vortex currents may be pinned.

The density of vortices was seen in Equation (8.1.1) to be proportional to the field strength, at least for $H \gg H_{c1}$. The pinning energy per unit volume is independent of field, however, so that the pinning force remains constant. Hence, if the vortices are to remain pinned,

$$J\sigma\phi_0 \leqslant \alpha_c , \qquad (8.1.12)$$

where α_c is a critical value.

If the current density and vortex density within the superconductor together satisfy Equation (8.1.12), no vortex movement will occur and no voltage gradient will arise. The superconductor will be truly resistanceless.

Equation (8.1.12) is found experimentally to be invalid very close to H_{c2}. It is also clearly invalid for $\sigma = 0$. This latter paradox is removed by setting

$$J \times (B + B_0) = \alpha_c , \tag{8.1.13}$$

where B_0 and α_c are arbitrary constants. The physical meaning of B_0 is not clear. α_c is the value of the Lorentz force which overcomes the pinning strength at a given temperature. Equation (8.1.13) is found to agree quite well with experimental results for $H_{c1} < \mathrm{H} < H_{c2}$.

8.2 Flux creep and the critical state†

The phenomenon of flux creep (flux flow mentioned previously is a limiting case of this) was first observed in experiments on the magnetic flux shielding and trapping characteristics of tubes of type II superconductors. In these experiments an external field is applied parallel to the axis of the cylinder. As vortices penetrate into the cylinder, shielding currents are set up in the bulk of the superconductor which inhibit further flux penetration. Dislocations pin the vortices against the Lorentz force of the shielding current so that locally the relation (8.1.13) between field and current holds, that is

$$J \times (B + B_0) = \text{constant}. \tag{8.2.1}$$

As the external field is increased still further, flux penetrates into the bore of the cylinder until at sufficiently high external field strengths the field strength within the bore becomes comparable with the external field. If now the external field is decreased, the internal field strength will follow it, always being, however, slightly higher owing to the current circulating in the cylinder. If during this phase the external field is held accurately constant and the internal field accurately monitored, as by nuclear magnetic resonance techniques, it is found that H_i slowly decreases. The time variation of H_i is logarithmic, that is, the rate at which H_i decreases becomes steadily smaller.

If we define a 'terminal' value of H_i to be that at which the rate of change of H_i becomes immeasurably small, we may characterise the variation of H_i as a function of H_e by means of a hysteresis diagram such as Figure 8.2.1. In order to measure a rate of decay of H_i, the external field would be first raised, then lowered, to reach the point A, for instance. The variation of H_i with time is then found to be of the form

$$H_i(t) = M - N \ln t , \tag{8.2.2}$$

where M and N are constants.

This logarithmic decrease of H_i indicates that flux is 'creeping' through the wall of the cylinder at a steadily decreasing rate. Indeed the flux creep rate quickly becomes so slow that very sensitive techniques are needed to detect it.

† Bean (1962); Kim *et al.* (1962).

The presence of flux creep suggests that the vortices normally pinned against movement by lattice defects are being 'shaken loose' from the pinning sites by thermal activity.

Experimentally it is observed that the pinning strength α of a type II superconductor decreases with increasing temperature. This is also explicable in terms of a thermally activated flux movement.

In Equation (8.1.11) the pinning energy of a normal region of radius ξ was given as

$$\Delta\epsilon = \tfrac{1}{2}\pi\xi^2\mu_0 H_c^2 \text{ joules per unit length of vortex.} \tag{8.2.3}$$

This is equivalent to a restraining force given by

$$F\xi = \Delta\epsilon , \tag{8.2.4}$$

since an amount of work $\Delta\epsilon$ must be done against the pinning force in moving the flux core a distance ξ away from the pinning centre.

Actually, the pinning energy will vary considerably from this somewhat idealised value. So, it is a concession to reality to consider an average pinning force

$$F_p = \tfrac{1}{2}P\pi\xi\mu_0 H_c^2 , \tag{8.2.5}$$

where P is an 'adjustable parameter'.

Consider the situation within the cylinder walls. If H_e is large, for instance at points A of Figure 8.2.1, the internal and external fields will be roughly equal and the flux density in the cylinder wall will be given

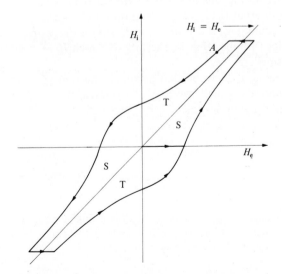

Figure 8.2.1. The variation of the magnetic field strength H_i within a tube of hard, type II superconductor as a function of the external field H_e.

approximately by

$$B = \tfrac{1}{2}\mu_0(H_e + H_i) \approx \mu_0 H_e \approx \mu_0 H_i \ . \tag{8.2.6}$$

The Lorentz force on one vortex in the presence of a transport current density J is $f_L = J\phi_0$ per unit length of vortex. Now the number of vortices per unit of area normal to the flux lines is given by

$$n_f = \frac{B}{\phi_0} \ . \tag{8.2.7}$$

If the average spacing between pinning centres is δ, then the number of centres per unit area is

$$n_p = \frac{1}{\delta^2} \ . \tag{8.2.8}$$

Thus the number of vortices per pinning centre is

$$\frac{n_f}{n_p} = \frac{B\delta^2}{\phi_0} \tag{8.2.9}$$

and the Lorentz force on the vortices per pinning centre and per unit length is given by

$$F_L = J \times B\delta^2 = \alpha\delta^2 \ . \tag{8.2.10}$$

If we combine (8.2.4) and (8.2.10), the net force on the vortex preventing movement is

$$F_n = F_p - F_L$$
$$= \tfrac{1}{2}P\pi\xi\mu_0 H_c^2 - \alpha\delta^2 \ . \tag{8.2.11}$$

Equation (8.2.11) as it stands implies that the pinning force is constant, unmodified by thermal vibration of the lattice. This is not so: the vibration of the lattice and therefore of a pinning centre itself will produce a random variation in the pinning force. By the random nature of this variation, the net pinning force F_p will occasionally be reduced to a level lower than the local Lorentz force on a group of vortices pinned by a defect. That is

$$F_n \leqslant 0 \ .$$

When this condition occurs, the vortices will be unpinned and will then flow under the action of the Lorentz force until the group is caught on another pinning centre. The frequency with which the vortices are unpinned will be a function of the net force F_n (ignoring thermal vibration) and of the temperature T.
    Thus,

$$R = W_0 \exp\left(-\frac{\xi F_n}{kT}\right) , \tag{8.2.12}$$

where R is the de-pinning frequency or flux creep rate, F_n is the net pinning energy, T is the absolute temperature, K is Boltzmann's constant, and W_0 is a characteristic frequency of vibration of the lattice, $\sim 10^{10}$ s^{-1}.

From Equation (8.2.11) it is seen that a slight reduction in F_L can produce a rapid rise in F_n. This in turn will considerably reduce the unpinning or creep rate R of Equation (8.2.12). Thus, a very small change in either J or B will effect a big change in creep rate.

However, it is now clear that bulk transport current in a type II super-conductor is never completely dissipationless: flux creep occurs at all levels of J and B although for small values of $J \times B$ it is unobservably small.

Suppose that, at given values of B and J, the resistance in a super-conductor due to flux creep is just detectable. Let this be a critical value of $J \times B = \alpha_c$ and let the corresponding critical value of $F_n = F_c$. Substituting this into Equation (8.2.12), we obtain

$$R_c = W_0 \exp\left(-\frac{\xi F_c}{kT}\right) . \tag{8.2.13}$$

that is

$$\xi F_c = kT \ln\left(\frac{W_0}{R_c}\right) . \tag{8.2.14}$$

If this value is substituted into Equation (8.2.11) for F_n, then

$$\alpha_c \delta^2 = \tfrac{1}{2} P \pi \xi \mu_0 H_c^2 - \frac{kT}{\xi} \ln\left(\frac{W_0}{R_c}\right) \tag{8.2.15}$$

(remembering that R_c is always less than W_0). Now from Equation (2.2.1)

$$H_c = H_0\left[1 - \left(\frac{T}{T_c}\right)^2\right] . \tag{8.2.16}$$

Inserting this in Equation (8.2.15) and collecting constants, we find

$$\alpha_c = \tfrac{1}{2} P \pi \mu_0 H_0^2 \frac{\xi}{\delta^2}\left[1 - \left(\frac{T}{T_c}\right)^2\right]^2 - \frac{kT}{\xi \delta^2} \ln\left(\frac{W_0}{R_c}\right) . \tag{8.2.17}$$

The form of Equation (8.2.17) is in close agreement with values of $\alpha_c(T)$ experimentally determined from many critical-state experiments. It infers that α_c falls to zero at a temperature below T_c. In fact, α_c remains finite up to T_c but Equation (8.2.17) is valid over most of the temperature range. A typical curve of α_c as a function of T is shown in Figure 8.2.2.

The other result to be obtained from this flux creep theory is the logarithmic decay of field with time found in tube magnetisation experiments.

Consider a tube of type II superconductor of large internal bore and small wall thickness having an internal field H_i and in an external field H_e.

The following conditions are assumed:

$H_i > H_e$,

$H_i + H_e \gg H_i - H_e$,

H_e is constant.

The current density J in the tube wall is constant across the wall. Then,

$$H_i - H_e = Jd \tag{8.2.18}$$

where d is the wall thickness. Also,

$$J \times (B + B_0) = \alpha , \tag{8.2.19}$$

where

$$B = \tfrac{1}{2}\mu_0 (H_i + H_e) . \tag{8.2.20}$$

The rate of decrease of field strength H_i depends on the creep rate R. Thus,

$$\frac{dH_i}{dt} = qRB , \tag{8.2.21}$$

where q is a constant depending on the geometry of the tube. From Equations (8.2.11) and (8.2.12)

$$R = W_0 \exp[-\xi(\tfrac{1}{2}P\pi\xi^2\mu_0 H_c^2 - \alpha\delta^2)(kT)^{-1}] \tag{8.2.22}$$

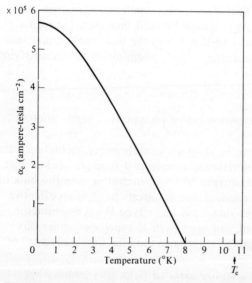

Figure 8.2.2. The variation of the effective pinning strength of a hard, type II super-conductor (niobium–titanium) as a function of absolute temperature.

and

$$\alpha = [\tfrac{1}{2}\mu_0(H_i+H_e)+B_0](H_i-H_e)d^{-1} . \qquad (8.2.23)$$

Since by assumption $H_i+H_e \gg H_i-H_e$, small changes in H_i will significantly affect only the second term in parentheses in Equation (8.2.23). Therefore α may be written

$$\alpha = rH_i-s , \qquad (8.2.24)$$

where r and s are constants. Hence Equation (8.2.22) becomes

$$R = W_0 \exp(-f+gH_i) \qquad (8.2.25)$$

where f and g are constants, and

$$\frac{dH_i}{dt} = \tfrac{1}{2} qW_0(H_i+H_e)\exp(-f+gH_i) . \qquad (8.2.26)$$

If we remember that H_i+H_e is treated as a constant, this expression integrates to give

$$H_i(t) = M-N\ln t , \qquad (8.2.27)$$

where M and N are constants. This is the Equation (8.2.2).

The term 'critical state' as applied to a hard, type II superconductor can now be better understood. A local region of a superconductor is said to be in a critical state when the local value of $\alpha = J \times (B+B_0)$ achieves the critical value α_c, such that flux creep proceeds at an arbitrarily slow rate. The application of a magnetic field to a hard, type II superconductor results in the creation of a critical state in the superconductor that extends in from the surface. As the external field increases, the region of critical states penetrates further. In this region the local current and flux densities will, through the mechanism of flux creep, attain values satisfying the relationship

$$J \times (B+B_0) = \alpha_c .$$

At this point, the essentially resistive nature of type II superconductors bears summarising.

The bulk of a type II superconductor is in a truly superconducting state because the energy of the conduction electrons in it is $\tfrac{1}{2}\mu_0 H_c^2$ below that of the normal state. However, quanta of flux penetrating into the bulk of the superconductor create an e.m.f. which manifests itself as an effective resistivity. Therefore whenever flux *moves* in a type II superconductor 'resistive' voltages are generated. In hard, type II superconductors flux movement is inhibited by pinning centres and the effective 'resistivity' of these materials is much lower than that of intrinsically soft, type II superconductors. Furthermore, in a steady external field, flux penetration into the hard, type II superconductor rapidly decreases and the associated 'resistivity' falls to an immeasurably small value.

8.3 Hard superconductors in the critical state†

The concept of the critical state can be used to predict the way in which magnetic flux will penetrate into a hard, type II superconductor and to explain the unstable magnetic behaviour of some superconductors used in magnet construction.

The first characteristic of hard superconductors to be successfully examined by means of the theory of critical states was their hysteretic magnetisation. A case of particular interest is the magnetisation of a layer of wires in a solenoid, for, as will be seen later, this may significantly affect the performance of the solenoid. This case must be approximated, for purposes of analysis, to that of a thin plate whose thickness is chosen in a manner described later.

Consider the sequential penetration of flux shown in Figure 8.3.1. Initially the plate is cooled below the critical temperature in the absence of a field. When ultimately at the temperature of the environment, the plate is subjected to a magnetic field. Up to the lower critical field H_{c1} of the plate, the field is totally excluded by surface currents, so that over this range of fields the magnetisation is given by

$$M = -H, \qquad H_{c1} > H > 0.$$

As soon as the field exceeds H_{c1}, flux moves inwards and the regions of

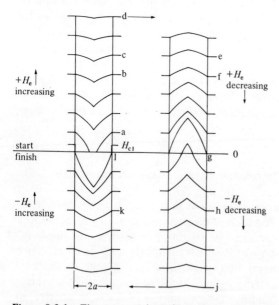

Figure 8.3.1. The sequential profiles of the penetration of a magnetic flux into a hard superconducting plate. The letters refer to points on the magnetisation curve (Figure 8.3.2).

† Bean (1962); Kim *et al.* (1962).

the superconductor immediately below the surface assume the mixed state with the formation of vortices. As the flux moves inwards through this region, it generates currents which are pinned by the mechanism already described. (Actually of course it is the vortices which are pinned against movement, but this is little more than a semantic point.) The rate of decrease of field strength in the mixed state region is given by

$$\frac{dH}{dx} = -J \ . \tag{8.3.1}$$

The local value of the current density J is as high as is allowed by the pinning strength. Thus,

$$J \times (B + B_0) = \alpha_c \ . \tag{8.3.2}$$

Combining Equations (8.3.1) and (8.3.2) and putting $B = \mu_0 H$, we get

$$\frac{dB}{dx} = -\frac{\mu_0 \alpha_c}{B + B_0} \ . \tag{8.3.3}$$

Integrating this expression and noting that, at $x = 0$, $B = \mu_0 H_e$, H_e being the external field strength, we obtain

$$\tfrac{1}{2}B^2 + BB_0 - \tfrac{1}{2}\mu_0^2 H_e^2 - \mu_0 H_e B_0 = -\mu_0 \alpha_c x \ , \tag{8.3.4}$$

whence

$$B = -B_0 + (B_0^2 + \mu_0^2 H_e^2 + 2\mu_0 H_e B_0 - 2\mu_0 \alpha_c x)^{\frac{1}{2}} \ . \tag{8.3.5}$$

The distance from the surface at which B/μ_0 has fallen to zero is given by

$$x^* = \frac{H_e}{\alpha_c}(\tfrac{1}{2}\mu_0 H_e + B_0) \ . \tag{8.3.6}$$

Now the magnetisation of this non-uniformly magnetised plate is defined by

$$M = \frac{1}{d_s}\int_0^{d_s} \frac{B}{\mu_0}dx - H_e \ , \tag{8.3.7}$$

where d_s is the semi-thickness of the plate. Using Equations (8.3.6) and (8.3.7), we may obtain an expression for the magnetisation of the plate, thus far in the cycle of applied field strength. Thus

$$\mu_0 M = \frac{1}{d_s}\int_0^{x^*} [-B_0 + (B_0^2 + \mu_0^2 H_e^2 + 2\mu_0 H_e B_0 - 2\mu_0 \alpha_c x)^{\frac{1}{2}}]dx - \mu_0 H_e \ . \tag{8.3.8}$$

At some external field strength H_e^* complete penetration of the plate occurs, with a central field strength equal to zero. This value of H_e^* is found from Equation (8.3.6) by substituting $x^* = d_s$:

$$\mu_0 H_e^* = -B_0 + (B_0^2 + 2\mu_0 \alpha_c d_s)^{\frac{1}{2}} \ . \tag{8.3.9}$$

At this juncture it is appropriate to consider the actual values of some of the parameters, for instance, H_e^*. Typically, a superconducting wire, such as would be used to wind a solenoid, might have the following parameters:

$d_s = 0.0125$ cm,

$\alpha_c = 10^6$ A T cm^{-2},

$B_0 = 1$ T,

hence $H_e^* = 8.25$ kA cm^{-1}. The lower critical field H_{c1} of most useful, hard, type II superconductors is typically about 400 A cm^{-1}. This is small compared with B_0 and, for conductors of a practical size, with H_e^*. Therefore we shall neglect it in the further consideration of the magnetisation cycle of the hard superconductor.

As the external field is raised above the value H_e^* required just to penetrate the plate, a series of flux penetration profiles is obtained, such as are shown in Figure 8.3.1. Because of the decreasing critical current density the slope of the profiles decreases, until at the upper critical field H_{c2} the magnetisation has fallen to zero.

The local current flow, induced by the movement of flux into the plate, is equal but of opposite sign on each side of the plate. The net transport current is thus zero.

For external fields exceeding H_e^* but not very close to H_{c2}, the magnetisation is given by

$$\mu_0 M_1 = -B_0 - \mu_0 H_e + \tfrac{2}{3}(B_0^2 + \mu_0^2 H_e^2 + 2\mu_0 H_e B_0 - 2\mu_0 \alpha_c d_s)^{\frac{1}{2}}(2\mu_0 \alpha_c d_s)^{-1}$$
$$-\tfrac{2}{3}(B_0^2 + \mu_0^2 H_e^2 + 2\mu_0 H_e B_0)^{\frac{1}{2}}(2\mu_0 \alpha_c d_s)^{-1}, \qquad (8.3.10)$$

where M_1 denotes the first quarter of the magnetisation cycle. This expression is plotted in Figure 8.3.2. For external fields approaching H_{c2}, α_c falls towards zero. Therefore the experimentally observed magnetisation decreases to zero at H_{c2}, whilst the theoretical expression does not. The curve oabcd clearly represents a diamagnetism, for the field strength within the superconductor is less than the applied field.

If at some point in the first diamagnetic quarter cycle of flux penetration (shown as the profiles abcd of Figure 8.3.1), the field strength is reduced a set of profiles such as efghj of Figure 8.3.1 is obtained. The consequent change in magnetisation is shown as the curve efghj in Figure 8.3.2. This curve represents the paramagnetic state of the super-conductor, for in this case the net field within the superconductor exceeds the applied field. Indeed, if the applied field strength is reduced to zero, flux will remain 'frozen' into the superconductor. This residual magnetisation is the cause of the residual field found in the bore of a superconducting magnet after its excitation current has been reduced to zero.

By an analysis identical with that used above for the calculation of the diamagnetism, the paramagnetism of the hard superconductor is

$$\mu_0 M_2 = -B_0 - \mu_0 H_e + \tfrac{2}{3}(B_0^2 + \mu_0^2 H_e^2 + 2\mu_0 H_e B_0 - 2\mu_0 \alpha_c d_s)^{3/2}(2\mu_0 \alpha_c d_s)^{-1}$$
$$+ \tfrac{2}{3}(B_0^2 + \mu_0^2 H_e^2 + 2\mu_0 H_e B_0)^{3/2}(2\mu_0 \alpha_c d_s)^{-1} , \qquad (8.3.11)$$

The magnetisation cycle is completed by a cycle of reversed field and the complete loop of Figure 8.3.2. is obtained.

Minor magnetisation loops, such as is represented by the path cefb in Figure 8.3.2, may be obtained by cycling the applied field between values at which singular penetration profiles are obtained. Such profiles have no discontinuity of slope between the surface and the midpoint, neither in the diamagnetic nor in the paramagnetic states.

The complete curve of magnetisation is a hysteresis loop, for it plots the lag of internal field behind the applied field. Like the hysteresis loop of iron its area is proportional to the energy loss incurred in one cycle of applied field, starting and ending at one field strength.

This applies for complete or for minor hysteresis loops, and is of importance in the a.c. application of hard superconductors.

So far we have considered a superconducting plate in an external field, without transport current. If a transport current is present, the penetration profiles and the magnetisation are modified.

It will be realised that a transport current gives rise to unequal field strengths on each side of the plate. Indeed the analysis of this case may proceed without reference to a transport current as such, but by consideration alone of the flux penetration profiles due to the two

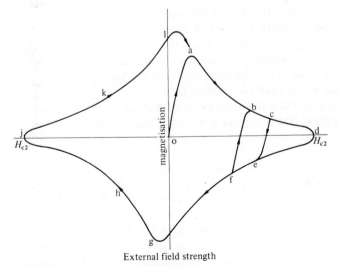

Figure 8.3.2. The sequential magnetisation of a hard superconducting plate. The letters correspond to the stages of flux penetration referred to in Figure 8.3.1. The path cefb is a minor hysteresis loop (see text).

unequal external fields. If a transport current density J_t flows the difference between the field strengths on each side of the plate will be

$$2\Delta H_e = 2d_s J_t . \tag{8.3.12}$$

Thus the applied strengths on each side of the plate are given by

$$H_{e1,2} = H_e \pm J_t d_s = H_e \pm \Delta H_e . \tag{8.3.13}$$

We are now considering the effects of the application of two fields, H_e and ΔH_e, which may be applied in different proportions to approach a particular end condition by an infinite possible number of paths. We shall briefly consider the first (diamagnetic) quadrant of the hysteresis loop for three paths: (a) H_e raised and held followed by ΔH_e, (b) ΔH_e raised and held followed by H_e, and (c) H_e and ΔH_e raised in the same proportions of their final levels. These three sequences of magnetisation are represented in Figure 8.3.3.

It is the last of these paths of magnetisation to which the conductors in a superconducting coil are subjected.

Sequence (a) is used in most tests of the critical current density of short samples of superconducting wire. The variation of the penetration profiles in this case is illustrated in Figure 8.3.4 whilst the corresponding magnetisation curve is shown in Figure 8.3.5. It will be noticed that on one side of the plate there is a discontinuity in the slope of the flux penetration profiles. Such a discontinuity produces stability on one side of the plate; this will be discussed later.

Sequence (b) is of little practical significance. It yields a profile without discontinuity.

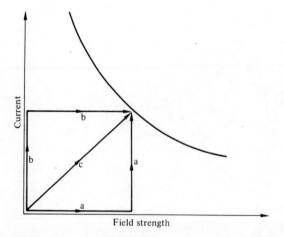

Figure 8.3.3. The three sequences of variation of field strength and current used in the tests of the critical current density of superconducting wires. In a the field is first raised to a set value followed by the current; in b the current is raised followed by the field; in c current and field are raised simultaneously.

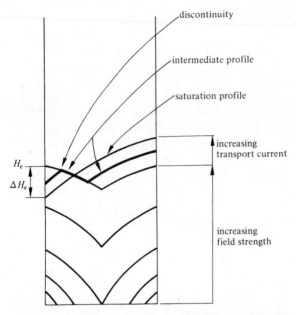

Figure 8.3.4. The variation of the flux penetration profiles in a plate (layer of wires) subjected first to a rising field and then to a rising current. The topmost profile is called the saturation profile. The current then flowing in the wire is the saturation current.

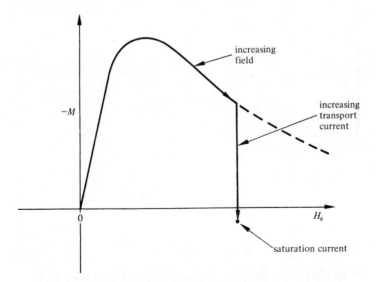

Figure 8.3.5. The paths followed along the diamagnetic branch of the magnetisation curve during the consecutive increase of field and current in the measurement of the critical current density of a superconducting plate (layer of wires).

Sequence (c) also yields a profile without discontinuity.

We shall examine sequence (a) in more detail.

As we shall see, a superconducting wire such as may be used for magnet construction is characterised by a short sample characteristic current, dependent on field (and temperature). Figure 8.3.6 shows the variation in the profiles of current density of a plate (or a layer of wires if we accept the analogy) with increasing transport current but with constant ambient field strength. The transport current increases the external field on one side of the plate and decreases it on the other. There is, however, no major change in the current density in the plate, which is everywhere the maximum allowed locally by the pinning strength. Thus the increase in transport current is achieved solely by the reversal of the local current flow close to the positions in the plate of the discontinuity of flux. As a result of this progressive reversal of part of

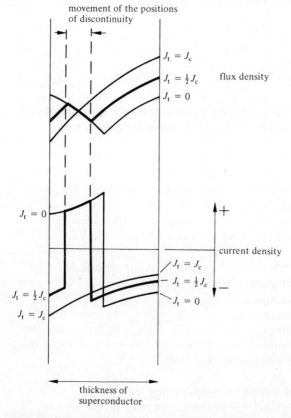

Figure 8.3.6. The variation in the profiles of current density and flux caused by the increase from zero of a transport current in an initially magnetised plate (layer of wires). This is the sequence of profiles obtained during tests of the critical current density of a plate or layer of wires. At saturation $J_t = J_c$.

the current flow in the plate the external field strength is increased on one side and decreased on the other. The limiting value of transport current, here termed the saturation current, is reached when all the current flow on one side of the plate has been reversed. At this point the flux penetration profile is as shown in Figure 8.3.6 extending continuously from the high external field on one side of the plate to the lower field on the other. Because of the dependence of the critical current density J on the local flux density through the relationship $J \times (B + B_0) = \alpha_c$, the flux profile at the saturation current is slightly convex upwards, thereby giving slight paramagnetism, at first sight a paradoxical phenomenon in view of the fundamental diamagnetism of superconductors.

The variation as a function of transport current of the magnetisation at constant external field is obtained easily if the critical current density is assumed to be independent of field strength. This approximation is valid for thin plates in which the variation of field strength is small. Then,

$$M = \frac{(J_t^2 - J_c^2)d_s}{J_c} , \tag{8.3.14}$$

where J_t is the transport current flowing per unit cross-sectional area of the plate.

Apart from the absence of a terminal paramagnetism at the saturation current, this parabolic variation of M with J_t agrees fairly well with experimental results on layers of thin wires.

It was mentioned at the outset that a layer of wires can be approximated by a thin plate of suitable chosen thickness and current density. In fact a good equivalence is obtained by making the thickness of the plate equal to the diameter of the wire. The effective current density in the plate is then chosen so as to give the same critical current per unit width of plate as is obtained from unit width of the layer of wires.

We have seen in this section how the concept of the critical state and of the local critical current density leads to the understanding of the pattern of flux penetration into a hard superconductor. In Section 9, armed with this basic understanding, we shall examine the reasons for the abrupt jumps in flux penetration that occur in most hard superconductors. But first we shall use the concept to examine what is generally called the 'critical current' of a hard superconducting wire.

8.4 The critical current of a hard superconducting wire†

In Section 8.3 we saw how flux penetrates into a hard superconductor by progressively creating locally the critical state. For ease of a well-defined analytical solution the critical state was assumed to be an exact and

† el Bindari (1969).

exact and unique relationship between flux and current density, given by

$$J \times (B + B_0) = \alpha_c , \qquad\qquad (8.4.1)$$

where α_c has a unique value.

In fact, however, the value chosen for α_c depends on what creep rate is accepted as being the 'critical' one. In this section we shall consider in more detail what happens in a superconducting wire which is everywhere in the critical state, such as is shown in the top profile in Figure 8.3.6.

An apparatus used for the measurement of the 'critical' creep rate of a hard superconductor is shown typically in Figure 8.4.1. We shall suppose that in such an apparatus we are measuring the voltage gradient along a wire as a function of increasing current at a number of fixed values of magnetic field. As the current rises, a series of penetration profiles such as those of Figure 8.3.6 are generated. As each new profile is created, flux creep readjusts them slightly with a rapidly decreasing rate of change. Thus with each increase in current in the wire a small voltage will be measured which will rapidly decay and, given long enough, will become immeasurably small. However, when at last the top saturation profile of Figure 8.3.6 is obtained, the voltage will not decay, for there is then no way in which the profile can change to reduce the Lorentz forces on the vortices. We shall call the current at which this profile is achieved the

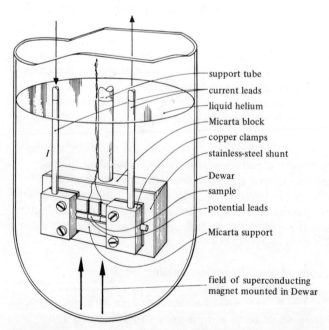

Figure 8.4.1. Typical apparatus for measuring the critical current density of a hard, type II superconducting wire.

saturation current. This current is referred to frequently as the short sample critical current.

The notation 'short sample' arose through a historical accident but is retained to differentiate from other 'critical' currents and current densities. Early in the construction of superconducting coils it was found that the current which would flow in a coil, before a catastrophic return of the normal state, was much lower than the current which would return normality to a short sample.

In those early days of superconducting magnets it was consequently assumed that wires of hard, type II superconductor could only be operated in coils at less than the 'short sample' critical current. Now we know in fact that they can even be operated above the saturation current. How this can occur is explained later.

Let us return to the sample of wire in which the saturation current is flowing. This is indicated by the presence along the length of the sample of an infinitesimal voltage which does not decay. Suppose the current is now increased very slightly. The Lorentz force on the vortices now exceeds the time-average pinning strength of the material and vortices move continuously across the superconducting wire, thus generating a steady voltage gradient along it. The motion of the vortices is governed partly by the frequency of thermal agitation of the pinning centres and partly by the 'viscous flow resistance' of the material, that is its flow resistivity. If the current is increased further, the motion of the vortices is dictated almost wholly by flux flow. Then the ratio of incremental voltage steps to the incremental current steps which cause them is the flow resistance. Figure 8.4.2 shows a typical experimental curve of voltage as a function of current for a hard, type II superconductor of very small cross section.

We see then that it is quite possible for a current exceeding the saturation current to flow in a hard, type II superconductor but accompanied by a

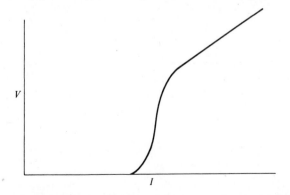

Figure 8.4.2. The voltage gradient along a hard superconducting wire as a function of the transport current in the wire at constant field strength, as measured typically.

voltage gradient. The voltage gradient is determined by the flow resistance. We shall simplify the processes occurring in this resistive region by assuming that at currents immediately above saturation the resistance takes on the value of flow resistance. Without any transition region then the variation of flow resistivity as a function of current at constant temperature is as shown in Figure 8.4.3.

If the temperature dependent saturation current is now termed I_c, the voltage gradient will be given by

$$E = (I - I_c)R_f ,$$ (8.4.2)

where I is the total current and R_f is the flow resistance per unit length.

Now, the voltage gradient interacts with the total current to generate heat given by

$$W = EI = I(I - I_c)R_f .$$ (8.4.3)

(Note that the heat generation is the product of voltage and *total* current.)

The heat generated will be discharged to the environment of the wire (for instance, liquid helium boiling at $4 \cdot 2°K$) causing a temperature rise in the wire. This temperature rise is given by

$$W = \gamma P(T - T_B) ,$$ (8.4.4)

where γ is the heat transfer coefficient, P is the wetted perimeter of the wire, T its temperature, and T_B the bath temperature. The rise in temperature reduces the pinning strength and hence the saturation current. It is a reasonable approximation to write the saturation current at a temperature T as

$$I_c = I_{cB} \frac{T_{cH} - T}{T_{cH} - T_B} ,$$ (8.4.5)

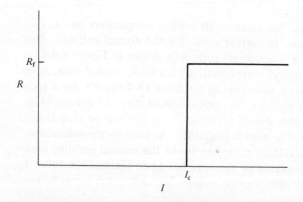

Figure 8.4.3. The assumed variation of the resistivity of a hard, type II superconductor as a function of current at constant field strength.

where I_{cB} is the saturation current at a temperature T_B and T_{cH} is the critical temperature of the superconductor in the prevailing ambient field strength.

With some algebraic manipulation Equations (8.4.2) and (8.4.5) combine to yield

$$v = \frac{i-1}{1-qi} , \qquad\qquad (8.4.6)$$

where

$$v = \frac{E}{I_{cB}R_f} ,$$

$$i = \frac{I}{I_{cB}} ,$$

$$q = I_{cB}^2 R_f [\gamma P(T_{cH} - T_B)]^{-1} ;$$

v, i, and q are non-dimensional parameters describing the voltage, current, and thermal insulation respectively.

Equation (8.4.6) reveals some interesting features of the resistive behaviour of hard superconductors.

The numerator shows that no voltage appears until the current exceeds the saturation current. Then the wire behaves resistively as if the excess current were flowing in the flow resistance of the wire. This is a concept used in consideration of the stability of hard superconductors. The denominator shows that the actual excess of transport current over saturation current depends on the temperature of the wire. Suppose that q is less than unity, which corresponds to a well-cooled wire. Then the voltage will increase stably with increasing current until the condition obtains that

$$v = i$$

At this point the wire has attained its critical temperature and is fully normal. Further increase in current generates the normal voltage. The variation of voltage as a function of current is shown in Figure 8.4.4.

If q is greater than unity, corresponding to a badly cooled wire, it is possible to obtain positive solutions to Equation (8.4.6) only for a range of values of i less than unity. The corresponding curve of voltage as a function of current is also shown in Figure 8.4.4. It can be seen that, if q is greater than unity, the wire is unstable. As soon as the saturation current is reached, the voltage rises directly to the normal resistive load line at which the temperature of the wire has again reached the value T_{cH}.

The stable operation of a wire at currents exceeding the saturation value is exploited in the technique of 'cryostatic stabilisation' described more fully in Section 9.4.

For the present we shall note how the property of some superconducting wires influences what is loosely termed as their critical current. It was shown that a hard superconducting wire is characterised by what is known as its short sample characteristic. As generally measured this is the current at which a given small voltage gradient is first observed, plotted as a function of field strength. Clearly in wires for which q is less than unity this 'short sample' current exceeds the saturation current. In wires of small diameter, coated with a normal metal of low resistivity such as copper, q can be considerably less than unity. Then the 'short sample' critical current may greatly exceed the saturation current if appreciable voltage gradients are used as the criterion for the reappearance of resistivity.

For example, consider a wire of diameter 0·025 cm having a flow resistivity of 3×10^{-6} Ω cm. Taking the heat transfer coefficient to liquid helium as 0·5 W cm^{-2} degK^{-1}, the critical temperature as 10°K in a field of 10 KA cm^{-1}, the bath temperature as 4·2°K, and the saturation current at 4·2°K as 50 A, then we find

$$q = 74 .$$

Such a wire would become fully normal at the saturation current, generating a large and unequivocal voltage gradient.

Suppose, however, that the same wire were coated with 0·0069 cm of high-conductivity copper of resistivity 10^{-8} Ω cm at 4·2°K. Then, replacing

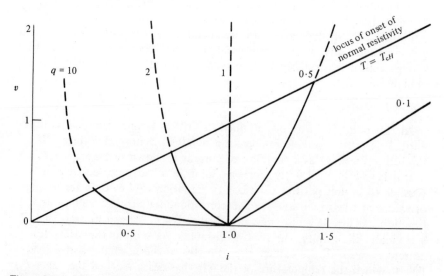

Figure 8.4.4. The predicted variation of voltage gradient along a hard, type II super-conducting wire as a function of current for various values of the thermal parameter q. The flow resistivity is assumed constant. The voltage rises until the normal resistive load line is reached, at which point the temperature of the wire is the critical value.

R_f by the net resistance per unit length, we obtain

$$q = 0 \cdot 1 \, .$$

Now stable operation is possible at currents above the saturation value. The onset of saturation is now not marked by a catastrophic return to the normal state but by the appearance of a small voltage gradient along the wire. Suppose that the sensitivity of measurement of this gradient is 100 μV cm^{-1}. This corresponds to a value of v of $0 \cdot 137$. From Figure 8.4.4 or Equation (8.4.6) for q equal to $0 \cdot 1$ this v gives a value of i of $1 \cdot 12$. Thus 56 amperes flows in the wire, roughly 6 amperes of this in the copper. This mode of current sharing can be a significant source of error in the measurement of the short sample characteristic of copper-clad wires. Current sharing is referred to again in more detail in connection with stabilisation (see Section 9.3).

In the following section we shall refer to the saturation current of a wire as its critical current, in conformity with general usage. It should, however, be borne in mind that short sample critical currents are meaningful only in relation to the q of the wire in its environment. The saturation current by contrast has a unique value at a given temperature and magnetic field.

Finally it must be pointed out that in deriving the curves of Figure 8.4.4 it was assumed that the flow resistivity was constant. In fact of course the flow resistivity is a function of temperature as well as of magnetic field strength. Thus in practice the curve of Figure 8.4.2 is obtained, in which the voltage increases smoothly towards that given by the normal resistance of the wire.

Superconducting magnets

9.1 Superconducting magnets

The discovery of hard superconductors has led to the large-scale exploitation of superconductivity in the construction of high-field magnets. A number of hard superconductors are now being used commercially. Their upper critical fields lie in the range 50 to 200 kA cm^{-1} and their pinning strengths are such as to allow current densities of 10^5 A cm^{-2} in this field range. Indeed it is in this application that superconductors are in greatest use. Consequently, the theory and practice underlying the construction of superconducting magnets will be considered in some depth in this chapter.

9.2 The magnetothermal instability of hard superconductors†

Were it not for an intrinsic mechanism of instability, hard superconductors could be used to wind coils almost as simply as copper wire is used to wind transformers. However, there exists an effect in hard superconductors which renders their performance in some types of coil rather unpredictable. Stated qualitatively the effect is this; the penetration of flux into a hard superconductor is essentially a dissipative process in which heat is generated. The generation of heat causes a temperature rise which, in accordance with Equation (8.2.17), decreases the pinning strength. This allows further flux penetration. Thus the penetration of flux is a regenerative process which may become catastrophic if certain conditions prevail. A catastrophic entry of flux into a hard superconductor is called a flux jump. What the conditions are which encourage the occurrence of a flux jump we shall now examine.

In Section 8.3 the equation for the profile of flux penetrating into a hard superconductor was found to be

$$ B = B_0 \left\{ \left[\left(1 + \frac{\mu_0 H_e}{B_0} \right)^2 - \frac{2\mu_0 \alpha_c x}{B_0^2} \right]^{\frac{1}{2}} - 1 \right\} . \tag{9.2.1}$$

The corresponding depth below the surface of the point of zero flux density is given by

$$ \delta = H_e (B_0 + \mu_0 H_c) \alpha_c^{-1} . \tag{9.2.2}$$

This is known as the shielding distance of a hard, type II superconductor. It must not be confused with the London penetration depth λ which is an intrinsic property of a superconductor.

In order to simplify the analysis it is necessary to make the modifying assumption that the current density is constant. Thus, instead of a critical pinning strength α_c we introduce a critical current density J_c, assuming a

† Hancox (1965); Bean et al. (1965); Swartz and Bean (1968).

constant field strength. Then Equation (9.2.1) becomes

$$B = \mu_0(H_e - J_c x) \tag{9.2.3}$$

and the penetration distance is given by

$$\delta = \frac{H_e}{J_c} . \tag{9.2.4}$$

These approximations are tantamount to the assumption that, in the expression $\alpha_c = J \times (B + B_0)$, B_0 is very large, or that the variation of B over the penetration distance is very small. In many cases these assumptions introduce negligible error. With these assumptions a typical penetration profile is shown as the solid line of Figure 9.2.1. Suppose that the external field strength H_e is momentarily raised by an amount ΔH_e. This will result in the flow of an amount of flux $\delta \mu_0 \Delta H_e$ per unit area at the surface. At a distance x below the surface the inflow of flux will be given by

$$\Delta\phi(x) = \mu_0 \Delta H_e(\delta - x) . \tag{9.2.5}$$

The effect of this amount of flux moving through a section of the super-conductor is to generate a voltage pulse of magnitude given by

$$e\,\mathrm{d}t = \mu_0 \Delta H_e(\delta - x) . \tag{9.2.6}$$

This voltage pulse interacts with the local current density J_c to generate a heat pulse of magnitude

$$\Delta q = J_c e\,\mathrm{d}t = \mu_0 J_c \Delta H_e(\delta - x) .$$

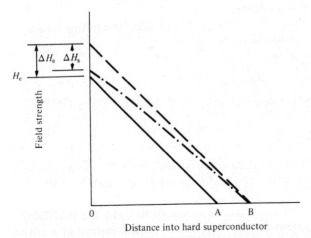

Figure 9.2.1. The penetration profile of magnetic flux in a semi-infinite block of a hard, type II superconductor. The solid line is the initial profile assuming constant temperature and current density. The dashed line is the profile after a small increase in external field. The chain profile includes the effect of locally increased temperature.

If the local volumetric specific heat of the superconductor is C, the local temperature will rise by

$$\Delta T(x) = \frac{\mu_0 J_c \Delta H_e}{C}(\delta - x) \ . \tag{9.2.7}$$

Now this rise in temperature will cause a decrease in the pinning strength so that, in effect, the local current density $J_c(x)$ will be reduced by an amount $\Delta J_c(x)$ given by

$$\Delta J_c(x) = \frac{\partial J_c}{\partial T}\Delta T(x) \ . \tag{9.2.8}$$

Thus,

$$\Delta J_c(x) = \frac{\mu_0 J_c \Delta H_e}{C}(\delta - x)\frac{\partial J_c}{\partial T} \ . \tag{9.2.9}$$

By integrating this decrease in current density over the shielding distance δ, we obtain an expression for the decrease in field strength which can now be shielded. Thus,

$$\Delta H_s = \int_0^\delta \Delta J \, dx \tag{9.2.10}$$

$$= \int_0^\delta \frac{\mu_0 J_c \Delta H_e}{C}(\delta - x)\frac{\partial J_c}{\partial T} \, dx \tag{9.2.11}$$

$$= \tfrac{1}{2}\delta^2 \frac{\mu_0 J_c \Delta H_e}{C}\frac{\partial J_c}{\partial T} \ . \tag{9.2.12}$$

Finally, substituting for δ from Equation (9.2.4) we get

$$\frac{\Delta H_s}{\Delta H_e} = \tfrac{1}{2}\frac{\mu_0 H_e^2}{C}\frac{\partial J_c}{\partial T} \ . \tag{9.2.13}$$

Now what this expression reveals is this. The increase in external field strength causes flux to enter the superconductor and extend the penetration distance from the point A to point B of Figure 9.2.1. If no decrease in current were to occur the dashed line would represent a new stable penetration profile. However, because of the increase in temperature due to flux movement the current density decreases. Thus, instead of this dashed profile, only that shown as a chain profile in Figure 9.2.1 is obtained. The increase ΔH_e, being transient, now disappears. The chain profile can then shield out the external field H_e within the distance 0B provided $\Delta H_s < \Delta H_e$.

Therefore, setting $\Delta H_s = \Delta H_e$, we obtain the limit of stability as

$$H_e^2 = 2J_c C\left(\mu_0 \frac{\partial J_c}{\partial T}\right)^{-1} \ . \tag{9.2.14}$$

The term $J_c(\partial J_c/\partial T)^{-1}$ has the dimensions of temperature. Instability results directly from the negative value of $\partial J_c/\partial T$ displayed by most hard superconductors. If $\partial J_c/\partial T$ could be made positive, inherent stability would result. This has been achieved in certain alloys.

A positive $\partial J_c/\partial T$ implies that, as temperature rises, pinning strength increases. This is in direct conflict with the model of flux pinning developed in Section 8.1. However, in that section we considered only the dependence on temperature of the pinning strength of one defect. If the number of defects available for pinning rises with temperature, even though their individual pinning strengths decrease, $\partial J_c/\partial T$ will be positive.

This has been accomplished in the alloy system Pb–In in which finely dispersed tin precipitate acts as the pinning agent. As the temperature rises, in the region of $3 \cdot 8°$K, the tin becomes normal and the particles act as pinning centres. Thus, over a somewhat limited temperature range $\partial J_c/\partial T$ is positive and the alloy is free of magnetothermal instability.

For the types of hard superconductor used in the construction of magnets, the variation of critical current density with temperature is reasonably approximated by a straight line on which zero critical current density occurs at the critical temperature at the given field strength. Then the approximation

$$J_c\left(\frac{\partial J_c}{\partial T}\right)^{-1} = T_{cH} - T_B$$

(where T_B is the bath or base-operating temperature) is valid. T_{cH} is the critical temperature at the given field strength and J_c is the critical current density at the temperature T_B. With this approximation Equation (9.2.14) becomes

$$H_e^2 = \frac{2C}{\mu_0}(T_{cH} - T_B) \ . \tag{9.2.15}$$

From this it is concluded that the field at which instability first occurs in a semi-infinite hard superconductor depends only on specific heat and critical temperature. For the maximum stably shielded field both of these parameters must be large.

The base-operating temperature T_B may, of course, be chosen anywhere in the range 0 to T_{cH} although it is usual to immerse superconducting magnets in liquid helium boiling at $4 \cdot 2°$K. If T_B is raised, the bracketed term of Equation (9.2.15) decreases. However, the specific heat C at first increases more rapidly than $T_{cH} - T_B$ decreases, so that the stably shielded field H_e rises to a maximum at a temperature lying typically between 7 and $11°$K, depending on the superconductor. Above this temperature the maximum stable field decreases. This typical variation of stably shielded field with base-operating temperature is shown in Figure 9.2.2.

It will be appreciated that the assumption that the increase in flux density is the same everywhere as the increase in external field strength implies that the superconductor behaves in an adiabatic fashion grossly, but not locally. This of course is not strictly true. However, if local adiabaticity is taken into account, the theory leads to essentially the same result.

What we have considered thus far is the maximum external field strength that can be stably and completely shielded from the interior of a thick superconductor; that is at some point within the superconductor the flux density falls to zero. We may invert this criterion to determine the maximum thickness of a plate (or, with licence, a round wire) in which a magnetothermal instability should not occur.

It is clear that the foregoing arguments apply for any thickness of superconductor provided only that the field strength just falls to zero at some point within it. Hence a superconducting plate of finite thickness, in which the limiting stable field as given by Equation (9.2.14) is just shielded out, is just stable. Now the distance at which the field is shielded out is given by

$$\delta = \frac{H_e}{J_c},$$

where J_c is the critical current density at the given field strength. Substituting for H_e from Equation (9.2.14), we obtain the criterion for the critical thickness of the plate; thus,

$$d_s^2 < 2C\left(\mu_0 J_{cH}\frac{\partial J_c}{\partial T}\right)^{-1}, \qquad (9.2.16)$$

where d_s is the maximum half-thickness of a stable hard superconducting plate.

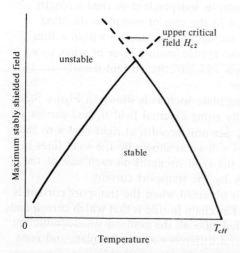

Figure 9.2.2. The variation of the maximum stably shielded field of a large hard superconductor as a function of temperature. The dashed curves are the extrapolation of the upper critical field and the stable shielding field.

Values for C, J_{cH}, and $\partial J_c/\partial T$ for a typical hard superconductor might be

$C = 10^{-3}$ J cm^{-3},
$J_{cH} = 3 \times 10^5$ A cm^{-2},
$\dfrac{\partial J_c}{\partial T} = 3 \times 10^4$ A cm^{-2} degK^{-1}.

These values lead to a critical half-thickness of $0\cdot0042$ cm. Although these criteria have been derived only by consideration of one-dimensional flux penetration into a block or plate, they are applicable to round wires without introducing errors greater, for instance, than that implicit in the assumption of constant specific heat. Thus the last criterion implies that the maximum diameter for a superconducting wire, of typical composition, in which instabilities will not occur is given approximately by

$$d_w^2 < 4C\left(\mu_0 J_{cH}\frac{\partial J_c}{\partial T}\right)^{-1}. \tag{9.2.17}$$

Because Equations (9.2.16) and (9.2.17) do not contain a term in field strength, they apply to any hard superconductor in which the flux density is attenuated towards the centre but not necessarily to zero.

We have thus far considered the way in which the specific heat capacity of a superconductor may prevent a flux jump from occurring. Next we consider how the deleterious effect of a flux jump, once initiated, can be limited by the thermal capacity of the superconductor.

9.3 Enthalpy stabilisation†
In order to reduce the analysis of this case to manageable form we shall again assume that the current density is independent of field strength. This introduces less error now than in the case of complete shielding because, as we shall see, the field strength (flux density) within a thin wire does not vary greatly. We also approximate a layer of wires to a homogeneous plate whose thickness and effective current density are chosen in a manner described later.

Consider a hard superconducting plate such as is shown in Figure 9.3.1. If this plate is situated in a steadily rising external field H_e and carries a steadily rising transport current I_t per unit breadth at right angles to the field, the flux penetration profiles will be as shown by the solid lines in Figure 9.3.1. The asymmetry in the field strengths on each side of the plate is, of course, caused entirely by the transport current.

Now this is the profile which is obtained when the transport current is less than the saturation current. The chain profile is that which corresponds to the saturation current, for, in that case, all the available superconduction current is used to maintain the field difference across the plate and none shields the field from the interior of the plate.

† Hancox (1966a).

This profile may be developed from that of the subcritical current merely by increasing the transport current. It may also be generated, however, by increasing the temperature. In this latter case, as the temperature rises, the current density decreases (because the pinning strength decreases) so that the slope of the flux penetration profile becomes less steep, this latter occurs because $dH/dx = -J_c$ when the temperature reaches the critical value (corresponding to the prevailing transport current and field strength).

In changing from the solid line (shielding profile) to the chain line (saturation profile) the penetrating flux interacts with the superconducting currents to generate heat. Suppose the shielding current density changes from J_c to $J_c - dJ_c$ everywhere within the plate. Then the penetration profiles will change as shown in Figure 9.3.2. At a distance x below the surface the amount of flux flowing across a line of unit length parallel to the surface is given by

$$\phi(x) = \int_x^\delta dB(x)dx \; , \tag{9.3.1}$$

where δ is the depth below the surface of the point of minimum flux density and where $dB(x)$ is the local increase in flux density due to the widespread fall in J_c.

This flow of flux, which has the units of volt seconds per centimetre interacts with the local current density J_c to produce a quantity of heat given by

$$q(x) = \phi(x)J_c \; . \tag{9.3.2}$$

The total quantity of heat liberated across the entire thickness of the plate

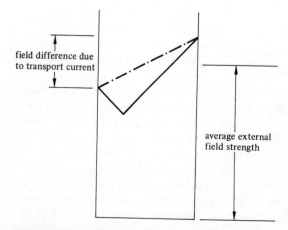

field difference due
to transport current

average external
field strength

Figure 9.3.1. The penetration profiles of magnetic flux in a plate (layer of wires) at the beginning and end of a magnetothermal instability. The transport current does not change, but the temperature rises, so that it becomes the saturation current at a higher temperature.

by the change dJ in current density is then

$$Q = \int_0^\delta \left[\int_x^\delta dB(x)\,dx \right] J_c\,dx \;, \tag{9.3.3}$$

in which the integration is performed for both sides of the plate. The heat released per unit volume of the plate is Q/D.

It is now necessary to substitute for dB and δ in terms of J_c. δ is determined by considering the value of the minimum flux density (field strength) within the plate. Thus,

$$H_{min} = H_1 - J_c\delta$$
$$= H_2 - J_c(D-\delta)\,, \tag{9.3.4}$$

whence,

$$\delta = \tfrac{1}{2}D \pm \frac{\tfrac{1}{2}(H_1 - H_2)}{J_c} \;. \tag{9.3.5}$$

$H_1 - H_2$ depends on the transport current in the plate and is given by

$$H_1 - H_2 = J_t \times D \;, \tag{9.3.6}$$

where J_t is the transport current density. Thus,

$$\delta = \tfrac{1}{2}D\left(1 \pm \frac{J_t}{J_c}\right). \tag{9.3.7}$$

From this expression it can be seen that a special case arises when $J_t = J_c$. Then, $\delta = D$ or 0 and the penetration profile is the single

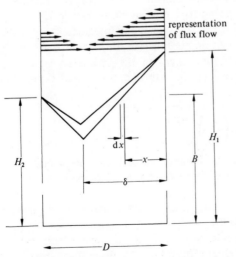

representation of flux flow

Figure 9.3.2. The change in flux penetration profiles caused by a reduction in critical current density.

straight line shown in Figure 9.3.1. J_t then corresponds to the critical current density.

dB is determined thus. At a depth x below the surface of the plate the flux density is given by

$$B(x) = \mu_0 H(x) = \mu_0(H_1 - Jx) . \qquad (9.3.8)$$

Therefore,

$$dB = -\mu_0 \, dJ_c .$$

With these substitutions Equation (9.3.3) becomes

$$\begin{aligned}
Q &= \frac{1}{D} \int_0^\delta \left[\int_x^\delta -\mu_0 x \, dJ_c \, dx \right] J_c \, dx \\
&= \frac{1}{D} \int_0^\delta [-\tfrac{1}{2}\mu_0(\delta^2 - x^2) dJ_c] J_c \, dx \\
&= -\frac{\mu_0 \delta^3 J_c \, dJ_c}{3D} .
\end{aligned} \qquad (9.3.9)$$

Substituting for δ, we find

$$\begin{aligned}
Q &= -\frac{\mu_0 J_c \, dJ_c \, D^2}{24} \left(1 \pm \frac{J_t}{J_c} \right)^3 \\
&= -\frac{\mu_0 D^2 J_c}{24} \left\{ 1 \pm \frac{3J_t}{J_c} + \frac{3J_t^2}{J_c^2} \pm \frac{J_t^3}{J_c^3} \right\} dJ_c .
\end{aligned} \qquad (9.3.10)$$

To obtain the total heat released during a change in critical current density from an initial value J_{c1}, corresponding to a starting temperature T_b, to a final value J_t at a higher temperature T_{cH}, we must integrate Equation (9.3.10) with respect to J_c. This must be done twice, for each side of the plate to which the alternate signs in the bracketed term correspond. Clearly the terms with dual sign cancel and the final expression for the total heat released becomes

$$Q_t = -\frac{\mu_0 D^2 J_{c1}^2}{24} \left\{ 1 - \frac{J_t^2}{J_{c1}^2} \left[1 + 6\ln\left(\frac{J_t}{J_{c1}}\right) \right] \right\} . \qquad (9.3.11)$$

Now what we have found so far is the amount of heat that would be liberated if the temperature of the superconductor were to be raised just to the critical value corresponding to a transport current $I_t = J_t \times D$ per unit width. Let us now find how much heat would actually be required just to achieve this critical condition.

Figure 9.3.3 shows a linear approximation of the variation of critical current (or critical current density) as a function of temperature for a given field strength. From this it can be easily seen that, at a temperature

T, the critical current density is given by

$$J_t = J_c \frac{T_{cH} - T}{T_{cH} - T_B} , \qquad (9.3.12)$$

where T_{cH} is the temperature at which the current is zero in the given field strength and where T_B is a base-operating temperature at which the critical current density is J_c.

If, for simplicity, we put $J_t/J_c = \beta$, then the temperature rise needed to reach the critical current density is $T - T_B$, which is given by

$$T - T_B = (T_{cH} - T_B)(1 - \beta) . \qquad (9.3.13)$$

The quantity of heat required to raise the temperature by this amount is

$$Q_P = \int_{T_B}^{(T_{cH} - T_B)(1 - \beta) + T_B} C(T)\,\mathrm{d}T . \qquad (9.3.14)$$

From Equation (9.3.11) the heat which can be released is

$$Q_T = \frac{\mu_0 D^2 J_c^2}{24}(1 - \beta^2 - 6\beta^2 \log\beta) . \qquad (9.3.15)$$

It can be seen that a value of β which equates the right-hand sides of Equations (9.3.14) and (9.3.15) is a value at which an instability will just bring the superconductor to its saturation profile. The bracketed term of Equation (9.3.15) is plotted in Figure 9.3.4.

Because the temperature dependence of the specific heat is a rapidly varying quantity at low temperatures, it is in practice necessary to consider

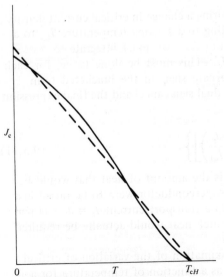

Figure 9.3.3. A linear approximation of the variation of critical current as a function of temperature for a given field strength.

the change in enthalpy required to raise the temperature to the critical value. Thus in Figure 9.3.4 may also be plotted the expression

$$q(\beta) = \frac{24}{\mu_0 D^2 J_c^2} \int_{T_B}^{(T_{cH} - T_B)(1-\beta) + T_B} C(T)\,dT \;, \tag{9.3.16}$$

where $\int_{T_B}^{T} C(T)\,dT$ is the enthalpy of the superconductor between the temperatures T and T_B.

The value of β at the intersection of the two curves gives the transport current density at which an instability can cause quenching.

In using Equation (9.3.16) to determine the degraded quenching current in a coil wound from wire of circular cross section, care must be taken in choosing an effective equivalent plate thickness D. Closest agreement between theory and practice is obtained when D is set equal to the diameter of the wire. However, close examination of this substitution

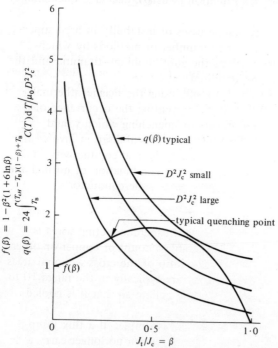

Figure 9.3.4. The heat liberated $f(\beta)$ and the heat absorbed $q(\beta)$ in a magnetothermal instability in a hard superconducting plate (layer of wires) plotted as a function of the parameter β. The value of β at the intersection of the curves defines a value of transport current (expressed as a fraction of the critical transport current at a base temperature) above which a magnetothermal instability will cause the wire to quench. $f(\beta)$ is an immutable function of β, but the curves of $q(\beta)$, the heat absorbed, depend on the characteristics of the wire. The three curves are only typical, showing the relative effect on stability of large, medium, and small values of $D^2 J_c^2$.

shows that the assumed plate exhibits greater diamagnetism than the layer of wires: hence more heat will be liberated per unit width and length of the assumed plate than in the wires. This is only partly offset by the greater thermal mass of the plate. Nevertheless, a good guide to the probable performance of a wire-wound coil can be obtained by this method.

Experimental verification of the fundamental truth of the method has been obtained by subjecting short samples of wire to 'pulse perturbation' tests (Iwasa and Montgomery, 1965). In these tests, the wire is carefully brought to a critical state at prescribed values of transport current and field strength; it is then exposed to a fast pulse of field of about 50 A cm^{-1}. This, of course, is the exact embodiment of the conditions hypothesised in the calculation of the stability limit of a semi-infinite superconductor. If at the point of pulsing, the transport current in the wire exceeds the value given by Equation (9.3.16), degraded quenching will occur.

We have analysed two fairly simple cases of instability in hard super-conductors. Now we shall examine a number of methods by which inherently unstable materials such as the alloy niobium–titanium and the compound niobium–tin can be stabilised.

There are basically two methods of stabilising the magnetothermal behaviour of hard superconductors; by preventing the occurrence of a flux jump or by preventing the process of quenching which would otherwise follow a flux jump. Both methods of stabilisation involve the use of normal material of high electrical and thermal conductivity in association with the superconductor. Such a combination of normal and superconductive material is called a composite superconductor.

9.4 Cryostatic stabilisation†

In this method of stabilisation one or more superconducting wires is formed into a composite with a considerable amount of copper or other normal metal of high conductivity. The ratio of the cross-sectional areas of the copper and of the superconductor are typically in the range 10 to 100. The composite is immersed in liquid helium so that it is cooled intimately by boiling heat transfer.

The principle of this type of stabilisation is simple: if a flux jump occurs in the superconductor so that this latter can no longer carry a superconduction current, the transport current will flow entirely in the copper. The heat which is then generated in the copper is dissipated to the liquid helium, with a concomitant temperature rise in the composite. If the temperature of the composite (and therefore also of the super-conductor) is less than the critical temperature of the superconductor in the prevailing conditions of field strength, the transport current will

† Kantrowitz and Stekly (1965).

return to it, thereby allowing superconducting operation to be resumed.
The simple condition for stability to which this model leads is that

$$I^2 \rho_n < a_n \gamma P(T_{cH} - T_B),$$ (9.4.1)

where I is the specified operating current, ρ_n is the resistivity of the
copper, γ is the heat transfer coefficient, a_n is the cross-sectional area of
the copper, P is the cooled perimeter of the composite, T_{cH} is the critical
temperature in the given field strength, and T_B is the temperature of the
helium bath (usually $4 \cdot 2°K$).

It will be borne in mind that T_{cH} is the temperature at which no superconduction current can flow in the superconductor. Current will,
however, commutate from the copper into the superconductor if this
process is mutually compatible with a fall in temperature. In fact this is
the case at all instants during the return of current to the superconductor
if Equation (9.4.1) holds. Furthermore, this expression is valid for all
values of I up to and including the saturation current, provided of course
that the other parameters of the criterion are appropriate.

Although simple, Equation (9.4.1) is only of limited applicability. It
neglects three important effects. The first of these is the nature of boiling
heat transfer to liquid helium. A curve of the heat transfer coefficient to
liquid helium as a function of temperature difference is shown in
Figure 9.4.1. This curve displays two discontinuous regions. In the lower
region, known as nucleate boiling, heat transfer coefficients as high as
$1 \cdot 0 \, \text{W cm}^{-2} \, \text{degK}^{-1}$ are obtained. At a temperature difference of about
1 degK this region ends, and what is known as film boiling commences.

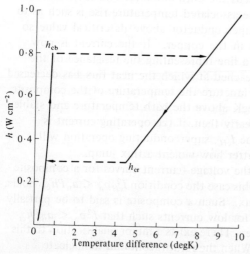

Figure 9.4.1. The typical rate of heat transfer to liquid helium boiling at 1 atm as a
function of temperature difference. The heat transfer is strongly dependent on the
geometry of the heated sample and this curve should be regarded only as an example.

At a heat flux of about 1 W cm^{-2} the temperature difference jumps to about 10 degK. If the heat flux is now reduced to 0·3 W cm^{-2}, an abrupt return to nucleate boiling occurs with a big drop in the temperature difference. The upward jump in temperature difference occurs at what is called the burn-out heat flux. the downward drop at the recovery heat flux.

The heat transfer coefficient γ in Equation (9.4.1) is thus seen to be a discontinuous function of temperature and, if the criterion is to be used at all, γ must be chosen with great care.

However, the expression can be modified thus

$$I^2 \rho_n < a_n P h_c ,$$

(9.4.2)

where h_c is a critical heat flux which may be chosen as either the burn-out or recovery value. If $I = I_{cH}$ (I_{cH} being the critical current in the given field) and if $h = h_{cr}$ (h_{cr} being the recovery heat flux), the composite is said to be fully stable. Under this condition, no matter how violent nor of what duration the flux jump, superconductive operation is always recovered for all currents up to the saturation value. The various degrees of cryostatic stability are illustrated by Figure 9.4.2 in which the voltage generated along a length of composite superconductor is plotted as a function of current.

In Figure 9.4.2a the condition $I_{cH}^2 \rho_n < a_n h_{cr}$ holds and the composite is fully stabilised. As the current is raised above I_{ch}, stable current sharing between superconductor and copper takes place and a voltage is generated along the composite. As the current is raised still further, the heat flux from the copper exceeds the burn-out value and the transition to film boiling takes place. The associated temperature rise is such as to raise the temperature of the superconductor above its critical value so that now all the current flows in the copper. If the current is now decreased, the voltage falls, on a line representing the resistance of the copper alone, until a point is reached at which the heat flux has decreased to the recovery value. At this juncture the temperature of the composite falls suddenly to less than 1 degK above the bath temperature and stable current sharing is resumed. Clearly then, if the operating current is restricted to the saturation value I_{cH}, superconducting operation will always be resumed after no matter how violent a flux jump.

In Figure 9.4.2b are traced the voltage–current curves for a composite of somewhat less stability. In this case the condition $I_{cH}^2 \rho_n < a_n P h_{cb}$ holds, h_{cb} being the burn-out heat flux. Such a composite is said to be partially stabilised, although, of course, for low currents such that $I^2 \rho_n < a_n P h_{cr}$ fully stable operation is possible. There exists a minor qualification to this degree of cryostatic stability. When the return from film to nucleate boiling occurs, a temperature difference still exists between the composite and its surroundings. If the temperature of the superconductor then remains above the critical value, superconductive operation cannot be recovered. However, the temperature difference after the fall from film to nucleate boiling is very small. Unless the superconductor is being operated at very low current near to its critical field strength, this qualification is unlikely to present a practical limitation.

There are, however, two other mechanisms which may prevent the complete return to superconductive operation following a flux jump. The first of these is the thermal barrier between the superconducting element of the composite and the copper.

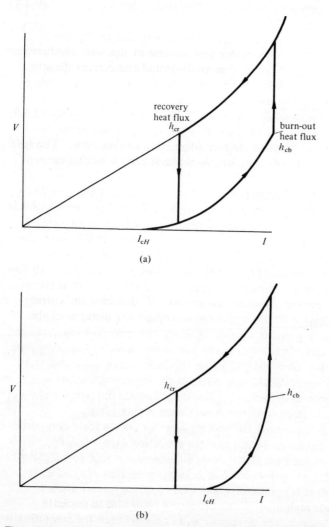

Figure 9.4.2. The voltage gradient along a cryostatically stabilised superconductor plotted as a function of transport current for various degrees of stability.

Consider a superconducting wire of diameter d_s embedded in a matrix of copper of cross-sectional area a_n and resistivity ρ_n. The heat transfer coefficient (thermal barrier) between the superconductor and the copper is γ W cm^{-2} degK^{-1}. We assume that the superconductor is carrying its saturation current and that the copper is at the temperature of the surroundings. Suppose that, as the result of a small temperature rise in the superconductor, a current ΔI_s transfers to the copper. This current produces a voltage gradient which is given per unit length by

$$v = \frac{\Delta I_s \rho_n}{a_n} = \frac{a_s}{a_n} \Delta J_c \rho_n \quad , \tag{9.4.3}$$

where ΔJ_c is the decrease in critical current density due to the rise in temperature. The heat generated per unit volume of the superconductor is given by the product of this voltage gradient and the current density in the superconductor. Thus,

$$w_s = v \times J_c = \rho_n J_c \Delta J_c \frac{a_s}{a_n} \quad . \tag{9.4.4}$$

The superconductor is of course in a resistive state of flux flow. The total rate of evolution of heat per unit length of the superconducting element is

$$W_s = \rho_n J_c \Delta J_c \frac{a_s^2}{a_n} \quad . \tag{9.4.5}$$

Now,

$$\Delta J_c = \frac{\partial J_c}{\partial T} \Delta T_1 \quad , \tag{9.4.6}$$

where ΔT_1 is the increase in temperature *required* to reduce the current density by ΔJ_c. The temperature rise *produced* by the heating w_s per unit length is

$$\Delta T_2 = W_s (\pi \gamma d_s)^{-1} \quad . \tag{9.4.7}$$

For stability $\Delta T_2 < \Delta T_1$, that is

$$W_s < \pi \gamma d_s \Delta T_1 \quad . \tag{9.4.8}$$

Thus,

$$\frac{\partial J_c}{\partial T} \frac{a_s^2}{a_n} \rho_n J_c \Delta T_1 < \Delta T_1 \pi d_s \gamma \quad . \tag{9.4.9}$$

This criterion, suitably recast, becomes

$$\tfrac{1}{4} J_c^2 d_s < \frac{a_n}{a_s} \frac{\gamma}{\rho_n} J_c \left(\frac{\partial J_c}{\partial T} \right)^{-1} \quad . \tag{9.4.10}$$

Typical values for the parameters of this expression are the following;

$$J_c = 10^5 \text{ A cm}^{-2}, \qquad\qquad \rho_n = 10^{-8} \text{ } \Omega \text{ cm},$$
$$d_s = 0.025 \text{ cm},$$
$$\frac{a_n}{a_s} = 10, \qquad\qquad J_c\left(\frac{\partial J_c}{\partial T}\right)^{-1} = 10 \text{ degK}.$$

Then, for stability, $\gamma > 0.006 \text{ W cm}^{-2} \text{ degK}^{-1}$. This is easily obtained in a co-drawn cryostatically stable composite (see Section 9.6).

It should be noted that Equation (9.4.10) applies for any number of superconducting strands in a matrix of normal metal. a_n/a_s is the ratio of total cross-sectional areas and d_s is the diameter of an individual super-conducting strand.

The second mechanism which may prevent the return of the composite to full superconductive operation concerns the thermal conductivity of the superconducting element: it is less easily analysed. Because the thermal conductivity of most hard, type II superconductors is very low, the heat generated in the superconductor by the mechanism just described will create a significant thermal gradient. Thus the temperature at the centre of the superconducting wire could be much greater than that of the matrix under resistive conditions. So, although the composite is cryostatically stable according to the criterion of Equation (9.4.2), the superconductor might be unable to carry the current because of an excessive temperature rise in part of it. The criterion for the maximum size of wire in which stability can be maintained against this effect is given by

$$J_c^2 d^2 \leqslant \tau \frac{K_s}{a_s}\left(\frac{a_n}{\rho_n} + \frac{a_s}{\rho_f}\right) . \tag{9.4.11}$$

This criterion is derived in the following section on dynamic stabilisation.

It will now be evident that the price paid for cryostatic stability is overall current density. The presence of a matrix of high-conductivity normal material reduces the effective current density in the composite as a whole. However, although limiting the overall current density, cryostatic stabilisation permits the use of very large currents in conductors of large section.

With the use of composites of very large section there arises, however, another form of instability which increases in severity as the size of the composite increases; it does not depend strongly on the size of the individual superconducting strands. Consider Figure 9.4.3 which shows a composite of rectangular cross section containing a number of super-conducting strands. For the sake of simplicity these are shown to lie at the edges of the composite. The conductor is assumed to be wound into the form of a disc coil in which the edges of the composite lie in the faces of the coil. Incident on the faces of the disc coil is a component of field $H_\perp$ which induces shielding currents in the superconducting strands at the edges of the composite. It is, furthermore, assumed that the field is changing with time. The dimensions of the composite conductor are given in the figure.

The currents induced in the strands of superconductor will maintain a difference in field strengths between the outside and the inside of the conductor. The magnitude of the shielding currents will be determined by the critical current density of the superconductor; thus,

$$I_s = J_c \times a_s , \qquad (9.4.12)$$

where a_s is the cross-sectional area of the superconductor at one edge of the strip.

Now these shielding currents must cross the composite strip over some distance in order to form closed current loops. The characteristic distance l_c measured along the length of the composite over which these shielding currents cross from one side to the other depends on the rate of change of the field $H_\perp$. Thus,

$$\mu_0 \left(\frac{\mathrm{d}H_\perp}{\mathrm{d}t} \right) = \frac{J_c a_s \rho_n}{l_c^2 t} , \qquad (9.4.13)$$

where ρ_n is the resistivity of the normal matrix (which may be copper) and where t is its thickness. The characteristic distance l_c may be quite short compared with the total length of conductor in a coil. For instance, enumerating the parameters as for the bubble chamber magnet described in Section 9.7, we have

$$\frac{\mathrm{d}H_\perp}{\mathrm{d}t} = 1 \text{ A cm}^{-1} \text{ s}^{-1}, \qquad \rho_n = 1 \cdot 32 \times 10^{-8} \ \Omega \text{ cm},$$

$$t = 0 \cdot 25 \text{ cm},$$

$$J_c = 10^5 \text{ A cm}^{-2}, \qquad a_s = 0 \cdot 017 \text{ cm}^2,$$

and then $l_c = 85$ cm. For a large magnet such as might use a composite

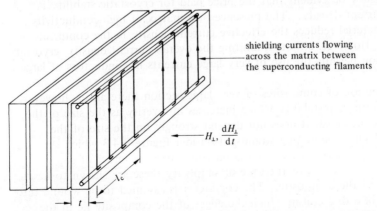

shielding currents flowing across the matrix between the superconducting filaments

$H_\perp, \dfrac{\mathrm{d}H_\perp}{\mathrm{d}t}$

l_c

t

Figure 9.4.3. A composite superconductor showing the paths of the shielding currents flowing in the superconducting strands at the edge of the composite and returning through the body of the composite at low current density.

of the assumed dimensions, this length is very short compared with the length of the conductor in the winding. Thus, even for slow charging rates the shielding currents flow exactly as if the whole composite were superconducting. In this case we may use a simple theory, similar to that of Section 9.2, to determine the quantity of heat generated if a flux jump should occur by which the internal and external field strengths of the composite are equalised.

The difference between internal and external field strengths is given by

$$H_{\perp e} - H_{\perp i} = \frac{J_c a_s}{t} .$$
(9.4.14)

Then, during complete flux penetration the flux entering the strip per unit length is $\phi = \mu_0 w(H_{\perp e} - H_{\perp i})$. This interacts with the shielding currents to generate heat given per unit length by

$$W = \mu_0 w I_s (H_{\perp e} - H_{\perp i})$$

$$= \frac{\mu_0 w J_c^2 a_s^2}{t} .$$
(9.4.15)

Substituting the values given above into Equation (9.4.15) with the width additionally set equal to 5 cm, we obtain

$$Q = 0 \cdot 725 \text{ J cm}^{-1}$$

and the heat density is

$$q = 0 \cdot 58 \text{ J cm}^{-3} .$$

This would cause a temperature rise in the copper of about 24 degK at which, of course, the superconductor would be normal. Furthermore, as may be inferred from Figure 9.4.1 the heat transfer rate at such a temperature difference might be sufficiently low to render the recovery of superconductivity impossible.

To remedy this situation it is necessary to reduce the shielding currents. This can be accomplished by arranging that the strands of superconductor are repeatedly transposed along the length of the composite, so that none lies on one side of the composite for a length greater than the characteristic length l_c. In this way insufficient voltage is developed in the transposition length by the changing field to maintain the flow of shielding current across the composite. Such transposition is difficult in conductors of high aspect ratio but is achieved in small wires to give an intrinsically stable superconductor. This will be described in Section 9.6.

At this juncture it may be useful to summarise the principles of cryostatic stabilisation. The presence of an environment of liquid helium is used to provide continuous cooling for a composite of superconducting strands embedded in a normal metal matrix of high conductivity, for instance, of copper. The term 'cryostatic' is used to denote a quasi-continuous process in which heat can be removed continuously from the composite. This type of stabilisation is not intended to prevent flux jumping. Indeed, for reasons noted above, particularly violent flux jumping may be observed in cryostatically stable composites. Good bonding between the superconductor and the copper is important; it ensures good heat transfer. The strands must be of less than a certain diameter to avoid appreciable temperature gradients within them. This critical size will be derived in the next section.

9.5 Dynamic stabilisation†

In the previous section we considered ways of alleviating the consequences of a catastrophic flux jump. Now we shall consider ways in which flux jumping can be prevented. We shall see that stabilisation designed only to prevent flux jumping incurs a far smaller decrease in overall current density than does cryostatic stabilisation.

The fundamental cause of instability in a hard superconductor is the generation of heat within it. If the rate at which this heat is generated is less than that at which it can be removed, flux jumping will be prevented. This is the principle of dynamic stabilisation, so called because the movement of flux is controlled.

Consider Figure 9.5.1 which shows the basic structure of a superconductive ribbon. A layer of hard superconductor of thickness d_s and critical current density J_c at a particular field strength is coated with a

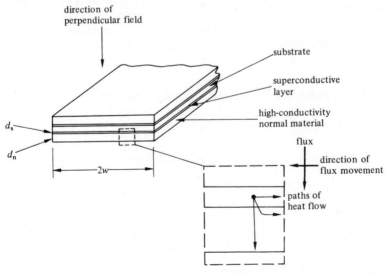

Figure 9.5.1. The structure of a typical superconductive ribbon: $d_s = 3 \times 10^{-4}$ cm, $d_n = 25 \times 10^{-4}$ cm, $w = 0.6$ cm, $K_s = 10^{-3}$ W cm^{-1} degK^{-1}, $K_n = 10$ W cm^{-1} degK^{-1}, $\rho_n = 10^{-7}$ Ω cm, $\rho_f = 30 \times 10^{-6}$ Ω cm, $\tau \approx 10°$K.

† Hart (1969); Chester (1967).

normal metal of resistivity ρ_n. It is assumed that the ribbon is everywhere in the critical state and that, because of a small temperature rise ΔT, flux flow is taking place. As a result of the temperature rise, the current flow is no longer resistanceless. A voltage gradient is produced as if a small part ΔI of the current were flowing in a resistive path.

Now,

$$\Delta I = 2wd_s \Delta T \frac{\partial J_c}{\partial T} . \tag{9.5.1}$$

This small current divides itself between the superconductor with flux-flow resistivity ρ_f and the normal material of resistivity ρ_n. The voltage gradient along the ribbon is given by

$$E = \Delta I \left(\frac{2d_s w}{\rho_f} + \frac{2d_n w}{\rho_n} \right)^{-1} = \Delta T \frac{\partial J_c}{\partial T} d_s \left(\frac{d_s}{\rho_f} + \frac{d_n}{\rho_n} \right)^{-1} . \tag{9.5.2}$$

This voltage interacts with the current in the superconductor (which is the main current in the ribbon) to generate heat within the superconductor given per unit volume by

$$W(T) = EJ_c = J_c \frac{\partial J_c}{\partial T} \Delta T d_s \left(\frac{d_s}{\rho_f} + \frac{d_n}{\rho_n} \right)^{-1} = f \Delta T . \tag{9.5.3}$$

Thus $W(T)$ is a function of the temperature rise ΔT. Now this heat generated in the superconductor must be removed by thermal conduction, either to the face of the strip or to its edge, depending upon its environment. We shall consider each of these cases in turn on the assumption first, however, that there is no thermal boundary resistance between the ribbon and its environment.

9.5.1 Edge cooling
The general expression for one-dimensional heat flow is

$$K \frac{\partial^2 T}{\partial x^2} - c \frac{\partial T}{\partial t} + W = 0 . \tag{9.5.4}$$

For T we may write ΔT and for W we may substitute Equation (9.5.3). Then,

$$K \frac{\partial^2 \Delta T}{\partial x^2} - c \frac{\partial \Delta T}{\partial t} + f \Delta T = 0 . \tag{9.5.5}$$

If x is measured from the centre of the ribbon, the general solution to Equation (9.5.5) is

$$\Delta T = \sum_0^\infty Mn e^{\alpha t} \cos \left[(2n+1) \frac{\pi x}{2W} \right] . \tag{9.5.6}$$

where n is an integer defining each of an infinite number of spatial temperature distributions. If we operate on Equation (9.5.5) according to the general solution (9.5.6) we find that a condition for there to be a solution is

$$\alpha c = f - K(2n+1)^2 \left(\frac{\pi}{2W}\right)^2 , \tag{9.5.7}$$

and the condition for stability is that all solutions for ΔT should decay with time.

Thus, $\alpha c < 0$ for all values of n is the required condition for dynamic stability in the strip. If $\alpha c < 0$ for $n = 0$ the condition will be satisfied for all values of n. The criterion for stability is

$$f < \frac{K\pi^2}{4w^2} . \tag{9.5.8}$$

Now the constant f was derived as if all the heat were generated within the superconductor. For most purposes this is correct; however, the heat can be conducted to the edge of the ribbon through both the superconductor and normal metal. Therefore the effective thermal conductivity K must be referred to the superconductor; it is given by

$$K = K_s + K_n \frac{d_n}{d_s} .$$

Our stability criterion therefore becomes

$$J_c \left(\frac{\partial J_c}{\partial T}\right) d_s \left(\frac{d_s}{\rho_f} + \frac{d_n}{\rho_n}\right) < \frac{\pi^2}{4w^2} \left(K_s + K_n \frac{d_n}{d_s}\right) . \tag{9.5.9}$$

For algebraic simplification we write

$$J_c \left(\frac{\partial J_c}{\partial T}\right)^{-1} = \tau, \text{ a characteristic temperature.}$$

Then,

$$J_{cs}^2 w^2 < \frac{\pi^2}{4} \tau \frac{K_s + K_n(d_n/d_s)}{d_s} \left(\frac{d_n}{\rho_n} + \frac{d_s}{\rho_f}\right) . \tag{9.5.10}$$

Now what this criterion gives is a limiting value for the critical current density, synonymous with pinning strength, in a given field. If the critical current density is less than this value, any thermal disturbance will decay. The term J_{cs} does not refer to a maximum value of transport current density that may be carried stably: it refers to the intrinsic pinning strength of the superconductor and is the maximum critical current density for which an instability will not occur.

By way of example we shall consider the ribbon illustrated in Figure 9.5.1 with the following values assigned to the parameters:

$d_s = 3 \times 10^{-4}$ cm,	$K_n = 10$ W cm degK^{-1},
$d_n = 25 \times 10^{-4}$ cm,	$\rho_f = 3 \times 10^{-6}$ Ω cm,
$w = 0\cdot6$ cm,	$\rho_n = 10^{-7}$ Ω cm,
$K_s = 10^{-3}$ W cm degK^{-1},	$\tau = 10°$K (typically).

If there were no high-conductivity cladding, edge cooling would limit the maximum permissible critical current density for stability to 48 A cm^{-2}, which is of course too small to be of any use in a superconducting magnet. However, if copper cladding of the 25 μm thickness assumed above is applied to the ribbon, the increase in thermal and electrical conductivities is so great that J_{cs} rises to $6 \cdot 9 \times 10^5$ A cm^{-2}.

9.5.2 Face cooling

In this case the heat generation term f will be exactly as before. But the heat conduction term will be $K_s \pi^2 / 4 d_s^2$, w in Equation (9.5.10) being replaced by d_s because the thermal resistance to heat flow now resides only in the thickness of the superconducting layer. The stability criterion is now

$$J_c \left(\frac{\partial J_c}{\partial T}\right) d_s \left(\frac{d_n}{\rho_n} + \frac{d_s}{\rho_f}\right)^{-1} < \frac{\pi^2 K_s}{4 d_s^2} ,$$

that is

$$J_{cs}^2 d_s^2 < \frac{\pi^2}{4} \tau \frac{K_s}{d_s}\left(\frac{d_n}{\rho_n} + \frac{d_s}{\rho_f}\right) . \tag{9.5.11}$$

For the ribbon of Figure 9.5.1 and assigning the same values for the parameters as above, without copper cladding, we find

$$J_{cs} = 9 \cdot 5 \times 10^4 \text{ A cm}^{-2} .$$

and, with the 25 μm cladding,

$$J_{cs} = 4 \cdot 8 \times 10^6 \text{ A cm}^{-2} .$$

There is apparently a strong case for the face cooling of superconductive ribbons. However, the transfer of heat from the face of the ribbon to the cooling medium may now become the factor limiting the maximum stable critical current density and this we shall consider later.

From these examples it can be seen that, for a given type of cooling, a copper-clad superconductive ribbon is more stable than an unclad ribbon and can carry a higher overall current density. As we shall see later, cladding with a normal material is a standard method of construction for superconductive ribbons using niobium–tin. It allows stable operation at overall current densities of 10^4 to 10^5 A cm^{-2} depending on the thickness of cladding.

Here it must be pointed out that the criteria for dynamic stability so far developed are apparently independent of field strength. In practice it is found that all commercial superconductors are more stable in high field strengths. This is a direct consequence of the decreased critical current density in high field strengths. The characteristic parameter τ also changes with field strength although not in a marked way. τ is approximately equivalent to the transition temperature at a given field strength and is largely independent of operating temperature. (It will be observed in Figure 8.2.2 that τ increases at low temperatures because $\partial J_c/\partial T$ decreases there.) Thus, as field strength increases the decrease in τ has an unstabilising effect. This is, however, outweighed by the stabilising influence of the decreasing critical current density.

We have considered only the effect of a field perpendicular to the face of the ribbon. Clearly, in a real configuration, components of field strength both perpendicular and parallel to the face of the ribbon will be present. We shall therefore now briefly consider the influence of the component parallel to the face.

Because in most niobium–tin ribbons the thickness of the super-conductive layer is very small (nearly always less than 50 μm) the parallel component of field penetrates fully, being only slightly attenuated in the interior of the layer. Because the thickness of the layer is small, the amount of flux moving through the surface during flux penetration is also small. This means that little heating will occur in the superconductor and therefore that an instability will not occur as a consequence of this field component. Furthermore, the presence of the parallel field component reduces the critical current density, thereby exerting a stabilising effect. The criterion for the thickness of superconducting layer which will allow stable penetration of the parallel field component is given by Equation (9.2.16), that is

$$d_s^2 < 2c\left(\mu_0 J_c \frac{\partial J_c}{\partial T}\right)^{-1} .$$

In assessing the stability of a ribbon by means of any of the formulae for dynamic stability derived above, the vector sum of parallel and perpendicular field components must be used to determine the characteristic temperature τ and critical current density.

Equation (9.5.10) for the stability of an edge-cooled ribbon ignores the parallel field component except insofar as it reduces J_c. It assumes that the perpendicular component has fully penetrated the ribbon. If the latter state does not hold, a modified form of stability criterion applies. In this case a higher current density may be permissible because only the outer edges of the ribbon are affected. This effectively reduces the dimension w. Indeed w is now simply the penetration distance of the perpendicular field component. Although the permissible critical current density may now be higher, there is a limitation to the perpendicular field strength.

Consider a tightly wound disc coil. The effective mean critical current density is

$$J_m = J_c \times d_s(d_n + d_s + d_w)^{-1} , \tag{9.5.12}$$

where d_n and d_s are as before and d_w is the thickness of interturn insulation.

The distance of penetration of the perpendicular field component is then given by

$$\delta = \frac{H_\perp}{J_m} = \frac{H_\perp(d_n + d_s + d_w)}{J_c d_s} . \tag{9.5.13}$$

Substituting this value of δ for w in Equation (9.5.10), we get

$$H_\perp^2 < \frac{\pi^2}{4}\tau \frac{[K_s + K_n(d_n/d_s)]d_s}{(d_n + d_s + d_w)^2}\left(\frac{d_s}{\rho_f} + \frac{d_n}{\rho_n}\right) . \tag{9.5.14}$$

This criterion defines the maximum field that may be stably shielded from the edge of the ribbon. It is important in the design of disc-wound coils of niobium–tin ribbon, for it indicates the maximum radial field strength that can be tolerated in the plane of the discs, if edge cooling alone is used. It must, however always be applied in conjunction with Equation (9.5.10) and the condition $\delta < w$, that is

$$H_\perp(d_n + d_s) < wJ_c d_s . \tag{9.5.15}$$

We shall now consider the effect of the finite coefficient of heat transfer between the ribbon and its environment.

9.5.3 Surface thermal barrier

The derivation of the criterion for dynamic stability in the presence of a thermal surface barrier is quite simple. The general equation for the thermal balance in the ribbon following a small rise in temperature ΔT is obtained in a manner similar to Equation (9.4.10). However, we now consider the total heat generated per unit width and length of the ribbon rather than the local heat generation for it is that amount of heat that must be discharged from the face of the ribbon. Because of this the heat generation term now becomes $fd_s\Delta T$ [cf. Equation (9.5.3)] and the specific heat term becomes $cd(\partial\Delta T/\partial t)$, in which c is an average specific heat per unit volume for a ribbon of thickness d. The exact value of $c \times d$ is not important for it does not enter into the final criterion. The equation for the heat balance following a perturbation ΔT thus becomes

$$-cd\frac{\partial\Delta T}{\partial t} + fd_s\Delta T - h\Delta T = 0 , \tag{9.5.16}$$

where h is the coefficient of heat transfer from the surface to the environment. The condition for stability is $\partial\Delta T/\partial t < 0$ as before, which is equivalent to $fd_s < h$. Substituting for f as derived in Equation (9.5.3), we obtain

$$\frac{1}{J_c}\frac{\partial J_c}{\partial T}J_c^2 d_s^2\left(\frac{d_s}{\rho_f} + \frac{d_n}{\rho_n}\right)^{-1} < h . \tag{9.5.17}$$

The criterion for dynamic stability under conditions of limiting heat transfer at the face of a ribbon is thus

$$J_{cs}^2 d_s < \frac{\tau h}{d_s}\left(\frac{d_s}{\rho_f} + \frac{d_n}{\rho_n}\right) .$$ (9.5.18)

By way of example we again consider the copper-clad ribbon of Figure 9.5.1 and assign values to the parameters as before. Then inserting a value of 1 W cm^{-2} degK^{-1} for the heat transfer coefficient in Equation (9.5.18) gives

$$J_{cs} = 1 \cdot 67 \times 10^6 \text{ A cm}^{-2} .$$

For an edge-cooled ribbon the area from which heat can be discharged to the environment is reduced in the ratio $(d_s + d_n)/w$ and the limit of stability is reached when

$$J_{cs}^2 d_s < \frac{\tau h}{d_s}\left(\frac{d_s}{\rho_f} + \frac{d_n}{\rho_n}\right)\frac{d_s + d_n}{w} .$$ (9.5.19)

Applying this criterion to the copper-clad ribbon, it is found that

$$J_{cs} = 114 \times 10^3 \text{ A cm}^{-2} .$$

In the event that the actual critical current density of the superconductive layer is greater than this value, Equation (9.5.13) may be used to transform (9.5.19) to an expression for the maximum stably shielded perpendicular field. This field is then given by

$$H_\perp = \frac{\tau}{J_c}\frac{h}{\rho_n}\frac{d_n}{(d_s + d_n + d_w)}\frac{d_s + d_n}{d_s} .$$ (9.5.20)

where d_w is the thickness of passive elements of a disc coil such as insulation, spacers. Typically for the same copper-clad ribbon $H_\perp$ is 0·22 kA cm^{-1}, if we assume its critical current density at this field strength is about $2 \cdot 5 \times 10^6$ A cm^{-2} and d_w is taken as 15 μm. If the ribbon is face cooled, there is no limiting perpendicular field strength.

In this section we have derived most of the criteria for the dynamic stability of superconductive ribbons of high pinning strength. The behaviour of ribbon, and this means principally niobium–tin ribbon, in superconducting coils bears a close qualitative comparison to these criteria. However, the performance of a coil cannot be predicted with exactitude from these formulae. In view of the complexity of the processes occurring in a hard, type II superconductor, this is not to be wondered at. Agreement between theory and practice is found to the extent, for instance, that ribbon coils quench initially in the end sections where the radial field component is high. Furthermore, if the same coils were to be operated in superfluid helium, which has a very high thermal conductivity, they would quench at much higher currents; this would arise because of the realisation of effective face cooling of the ribbon in the regions of high radial field strength.

Coils wound from a single, copper-clad, niobium–titanium wire operate generally at levels well above those which would be predicted by the theory of dynamic stabilisation. Presumably in these coils processes other than those associated with dynamic stability or cryostatic stability are influencing the stability.

To end this section we must return to the conditions for the cryostatic stability of a composite. It was mentioned that three conditions existed for cryostatic stability in a composite. Two have been dealt with in Section 9.4; the third was deferred to this present section because its analysis is identical with that of dynamic stability limited by thermal conductivity. The condition that the temperature rise within the superconductor of a cryostatically stabilised composite should decrease following the transient flow of current in the normal matrix is identical with criterion (9.5.10), derived above. Hence the criterion for the return of current to the superconducting strands in a composite is given approximately by

$$J_c^2 d_w^2 < \pi^2 \tau \frac{K_s}{a_s}\left(\frac{a_s}{\rho_f} + \frac{a_n}{\rho_n}\right) , \qquad (9.5.21)$$

where a_s is the cross-sectional area of the superconducting strands and d_w is here the diameter of the wire. This must of course be supplemented by the criterion for the thermal stability of the matrix in its environment.

9.6 Materials for superconducting magnets

Since 1960 a number of materials, both alloys and intermetallic compounds, have been used for the construction of superconducting magnets. Only two of these are now widely used; these will be described in some detail. The history of the others will be briefly mentioned.

The first superconducting magnet of any practical significance was wound with hard-drawn niobium wire. Although pure elemental niobium has an upper critical field of only 1600 A cm^{-1}, hard-drawn commercial wire has a critical field of 8000 A cm^{-1}. The increase is due to the reduction in coherence length caused by impurities and by dislocations introduced during cold drawing. Furthermore, these dislocations and other lattice defects give the drawn wire a pinning strength which allows a critical current density of about 10^4 A cm^{-2} at a field strength of 4000 A cm^{-1}.

The alloy molybdenum–rhenium was also used in a few early superconducting coils. It has an upper critical field of 11 000 A cm^{-1} and can support a current density of 10^4 A cm^{-2} at a field of 8000 A cm^{-1}.

In 1960 the first really high-field type II superconductor was used in magnet construction. This was the alloy niobium–zirconium. With 25 at.% zirconium this alloy has an upper critical field at 4·2°K of 56 kA cm^{-1}. At 32 kA cm^{-1} its critical current density is 10^5 A cm^{-2}. This material found extensive application in magnet construction between

1960 and 1966. The most usual form for the material was a cold-drawn wire of 0·025 cm in diameter, electroplated with 0·0025 cm of high-purity copper. The copper was applied in the first instance to reduce the resistive voltages generated during the quenching of a coil. However, it also improved the performance of the coil by increasing the stability of the wire. The quality of the bond between the copper and the niobium–zirconium was an important factor in the performance of the wire in coils. Although in Section 9.5 we did not specifically consider the effect of a thermal boundary resistance between the copper and the super-conductor, it is as important for the stability of wires in small coils as it is in cryostatic stabilisation.

It was soon realised that the best bond would be obtained in a process of co-drawing, in which a true metallurgical bond would be created. The mechanical properties of niobium–zirconium preclude a co-drawing process with copper, but the alloy niobium–titanium lends itself well to such a process. In 1965 co-drawn composites of niobium–titanium in copper were produced. As expected, coils wound from this wire showed a high degree of stability.

9.6.1 Niobium–titanium†

The alloy of niobium with 44 wt.% of titanium has an upper critical field at 4·2°K of 88×10^3 A cm^{-1}. Before 1965 it was recognised as a potential magnet material. However, it has a very low specific heat so that, according to Equation (9.2.6), it should be less stable than niobium–zirconium. Indeed the first niobium–titanium coils wound with wire having a cladding of poorly bonded electroplated copper could only be operated in high ambient field strengths. This is of course explained by the much reduced critical current density in high fields.

The stability shown by the first coils wound from co-drawn niobium–titanium wire was due to the good bond between superconductor and copper and to a lesser extent to the very low resistivity of the copper. This latter arose from the greater purity of the copper, which was not contaminated by the process of electrolysis, and from the final heat treatment of the wire. Although intended primarily to obtain the desired pinning strength in the superconductor, this heat treatment also effectively anneals the copper.

Although the details of the production of co-drawn niobium–titanium–copper composite wires are obscured by commercial secrecy, the essentials of the process are as follows.

Niobium and titanium powders are mixed and arc melted in vacuum to form an ingot. Typical ingot weights are 10 to 100 lb. The ingot is then extruded into a slug of high-purity copper with which it forms a cold-welded bond. This stage of the process is the most important and upon it depends the stability of the finished wire. This composite billet is then

† Vetrano and Boom (1965); Berlincourt and Hake (1962).

swaged and drawn to form a wire or larger composite. Intermediate annealing may be used to assist the drawing operations. The finished composite may be a single wire of overall diameter 0·04 cm or a large cryostatically stable conductor consisting of a rectangular copper matrix containing a number of niobium–titanium wires. It may be an intrinsically stable wire consisting of a copper matrix of overall diameter 0·04 cm containing one hundred very fine superconducting filaments. It may even be a hollow composite through which a cooling fluid can be circulated. There are few limits, except for ingot weight, to the size of the composite or to the combinations of superconductor and copper. In all types of niobium–titanium composite it is the final heat treatment which decides the properties of the superconductor. During the stages of cold working the superconductor and the copper assume an extensive structure of dislocations and defects. In the niobium–titanium these are necessary for the formation of pinning centres; in the copper they are an embarrassment. However, the pinning strength of the superconductor as first formed is comparatively low and would not allow a useful current density at high field strengths. By heating the cold-worked niobium–titanium at between 300 and 400°C for 30 min, precipitation, probably of dissolved oxygen, occurs at the dislocations greatly increasing their pinning strength. At the same time this heat treatment anneals the copper, so decreasing its resistivity.

Co-drawn niobium–titanium–copper composites produced in this way are now used almost exclusively for the construction of superconducting coils generating fields of up to 70 kA cm^{-1}. The main characteristics and properties of niobium–titanium are given in Table 9.6.1. The current density as a function of field strength is shown typically in Figure 9.6.1.

The construction of coils from single-core wire is essentially a simple task.

Table 9.6.1. Properties of niobium–44 wt.% titanium.

$H_{c2}(0)$ (kA cm^{-1})	96
$H_c(0)$ (kA cm^{-1})	2·8[a]
$H_{c1}(0)$ (kA cm^{-1})	0·112[a]
T_c (°K)	10·5
κ	24[a]
α_c at 4·2°K, 16 kA cm^{-1} (A T cm^{-2})	6 × 10^5
K at 4·2°K (W cm^{-1} degK^{-1})	0·001[b]
ρ_n at 11°K (Ω cm)	3 × 10^{-5}
c at 4·2°K (J cm^{-3} degK^{-1})	10^{-3}
τ at 4·2°K, 16 kA cm^{-1} (°K)	~5
Modulus of elasticity at 300°K (kP mm^{-2})	9500
Yield strength at 300°K (kP mm^{-2})	70

[a] These values are extrapolated from data of Feitz and Webb (1967) using the formula of Shapoval (1962).
[b] Unpublished measurement of Montgomery *et al.*

The wire is usually insulated with polyvinyl acetate or similar varnish and may be wound on to a coil former like copper wire. Precautions must be taken to insulate adjacent layers sufficiently to withstand the voltages that are likely to be generated during quenching. Also copper foil is usually placed between layers to conduct heat away from the centre of the winding where it is generated by flux movement through the wire.

Superconducting coils are of course potentially capable of operation in the persistent mode. This is achieved by short-circuiting the ends of a winding after it has been energised. A thermal switch is the device by which the short circuit is achieved. It consists of a short length of super-conducting wire without copper cladding, wound into a non-inductive coil, and provided with a heater. This switch is connected between the ends of the winding. During the charging of the coil the heater is energised to raise the temperature of the switch above its critical temperature. Because the superconducting wire in the switch is without copper cladding, its resistance in the normal state is high. When the coil has been fully charged, the switch heater is de-energised, so allowing the switch to become superconducting. The charging current is then reduced to zero; as this is done, the coil current builds up in the switch. Because the switch wire is unclad copper, its stability is poor. It is therefore usual to employ thin wire in the switch. If the diameter of the wire is sufficiently small, the wire will be inherently stable.

Superconducting joints may be made between niobium–titanium wires by spot welding. These joints will not retain the full pinning strength of the parent wires and must be located in regions of relatively low field strength.

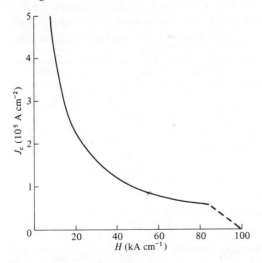

Figure 9.6.1. The variation of the critical current density of niobium–titanium as a function of field strength at $4 \cdot 2^\circ$K.

Single-core niobium–titanium wire is being used mainly in a size having a core of $0\cdot025$ cm and with a cladding thickness of $0\cdot004$ cm to $0\cdot012$ cm. We may examine the theoretical stability of these wires by invoking Equation (9.5.21). Although this criterion refers to the cryostatic stability of superconducting wires in a copper matrix, it is derived from the criteria for the dynamic stability of composites and is, in fact, the criterion also for the dynamic stability of the superconductor. We may therefore use Equation (9.5.21) in the present context of a lightly clad single niobium–titanium wire.

The typical critical current density of niobium–titanium in low field strengths is 2×10^5 A cm^{-2}. The thermal conductivity of the alloy at about $4\cdot2°$K is $0\cdot001$ W cm^{-1} degK^{-1} and the electrical resistivity of the copper cladding in low field strengths is about 10^{-8} Ω cm. If we insert these parameters in Equation (9.5.21), the required cross-sectional area of the cladding is $0\cdot013$ cm^2. This demands a radial thickness of $0\cdot053$ cm. Thus, the commonly used single-core niobium–titanium wires do not satisfy the criterion for dynamic stability, unlike the niobium–tin ribbons we shall be considering later.

In coils containing up to 20000 m of wire clad with $0\cdot0075$ cm of copper the current at which the coil quenches will vary between 50% and 100% of the saturation current. Flux jumping can be observed in most coils wound with this type of wire, further indicating that dynamic stability is not achieved in them. It is found that Equation (9.3.16) gives an approximate indication of the quenching current in these cases, confirming the lack of dynamic stability and suggesting that the quenching current is determined by the balance between the energy of magnetisation and the enthalpy of the wire. However, the mechanism of stabilisation in single-core wires of niobium–titanium is not completely explained by any of the theories we have considered and is indeed not yet fully understood. It probably involves the transient flow of current in the copper cladding, the maximum temperature during a flux jump being determined by the specific heat of the composite and by the resistance of the copper cladding.

A very important recent development in niobium–titanium composites is the production of a wire which is intrinsically stable.

Equation (9.2.17) is the criterion for intrinsic stability in a single wire. For niobium–titanium at $4\cdot2°$K and in low field strengths (where instability is greatest),

$J_{cH} = 3 \times 10^5$A cm^{-2} (in a field of 16 kA cm^{-1})

$\dfrac{\partial J_c}{\partial T} = 6 \times 10^4$A cm^{-2} degK^{-1} ($\tau \approx 5$)

$c = 10^{-3}$ J cm^{-3} degK^{-1}.

Substituting these values into Equation (9.2.17), we find

$d_w = 4\cdot2 \times 10^{-3}$ cm.

Thus, a niobium–titanium wire of less than about two-thousandths of an inch in diameter is inherently stable against flux jumping. To use a single wire of such diameter in the construction of a coil of useful size would of course be almost impossible. Furthermore, the manufacturing difficulties involved in making the single length required would be almost insurmountable. However a composite wire is made in which a large number (often 61) of niobium–titanium filaments of such diameter are co-drawn in a copper or other matrix. The final overall diameter may then be about $0·025$ cm, as for a typical single-core wire. This composite will contain about 40% by volume of normal metal matrix and 60% of superconductor. The diameter of the niobium–titanium filaments is about $0·002$ cm. The very small diameter of these filaments is a necessary but not a sufficient condition for magnetothermal stability. In the presence of time-varying fields, shielding currents will flow along the filaments on one side of the conductor, through the matrix, and back along the other side to form a shielding loop current of long decay time (see section 9.4). In order that these shielding currents should decay in a time short compared with the rise time of the field, the filaments must be transposed or at least twisted inside the matrix. Transposition, although ideal, is impracticable for any configuration of filaments other than a single circular layer. However, the filaments can be twisted fairly easily; actually, of course, the whole wire is twisted so that the applied field periodically reverses its orientation with respect to the filaments. The required pitch of this twist is given by the characteristic length l_c in Equation (9.4.13). In this expression t is interpreted as the diameter of the wire. A typical pitch for the twist is obtained, by way of example, by substituting the following values of the relevant parameters:

$$a_s = 0·000\,192 \text{ cm}^2,$$
$$t = 0·025 \text{ cm},$$
$$\rho_n = 10^{-7} \text{ } \Omega \text{ cm},$$
$$J_c = 3 \times 10^5 \text{ A cm}^{-2},$$
$$\frac{\mathrm{d}H}{\mathrm{d}t} = 800 \text{ A cm}^{-1} \text{ s}^{-1} \text{ (typically)}.$$

Then $l_c = 4·8$ cm. A twist of about one turn per centimetre is usually applied to filamentary composites of this kind.

Materials other than copper may be used for the normal matrix, whose electrical conductivity is not of importance in the stability of this type of composite. For instance, cupro-nickel, whose mechanical properties match those of niobium–titanium is also used as matrix material. Its use permits the operation of an intrinsically stable superconductor at very rapid rates of change of field.

Because the resistivity of cupro-nickel is much higher than that of copper, a given pitch of twist will allow a faster rate of rise of field. In fact, if in the previous example the matrix had been cupro-nickel ($\rho_n = 10^{-5}$ Ω cm), a rate of rise of field of 8000 A cm^{-1} s^{-1} would have resulted in the same level of shielding current in the wire. The advantage of a hard super-conductor which is intrinsically stable in rapidly changing fields lies in its application to a.c. systems, as we shall see in Section 10.2.

Coils are now constructed routinely with both the single-core copper-clad wire and the intrinsically stable fine wire. The single-core material operates in coils at currents up to the saturation level depending on the amount of copper cladding. The intrinsically stable wire operates in coils at its saturation current for all but the most rapid rates of rise of field.

A variety of typical composite niobium–titanium superconductors is shown in Plate II.

9.6.2 Niobium-tin†

Niobium–tin (Nb$_3$Sn) is an intermetallic compound having a β-tungsten structure, one which is found to favour superconductivity in compounds. The main characteristics of niobium–tin are set out in Table 9.6.2. It has but one disadvantage as a material for coil construction; it is exceedingly hard and brittle. Several ways have been devised to mitigate the problems resulting from this brittleness. Only two of these methods survive for the preparation of commercially available conductors and these alone will be described.

1. *Vapour deposition.* The problems associated with the brittleness of niobium–tin can in general be avoided by using very thin sections of the material. Just as glass can be bent if it is in the form of thin fibres, so can niobium–tin, whose mechanical properties are in many ways similar to those of glass.

In the vapour deposition technique a thin layer of the compound Nb$_3$Sn is chemically formed on a ribbon of stainless steel. In a typical commercial process a 'Hastelloy' tape of 0·0025 cm to 0·005 cm thickness and of width between 0·25 cm and 1·27 cm is continuously fed through a reaction chamber where it is heated to about 950°C by an electric current.

Table 9.6.2. Properties of Nb$_3$Sn.

$H_{c2}(0)$ (kA cm^{-1})	185
$H_c(0)$ (kA cm^{-1})	2·8
$H_{c1}(0)$ (kA cm^{-1})	0·18
T_c (°K)	18·3
κ	45
α_c at 4·2°K, 40 kA cm^{-1} (A T cm^{-2})	3×10^6
K at 4·2°K (W cm^{-1} degK^{-1})	0·5
ρ_n at 18·3°K (Ω cm)	10^{-5}
c at 4·2°K, zero field (J cm^{-3} degK^{-1})	$4·5 \times 10^{-3}$
τ at 4·2°K, 40 kA cm^{-1} (°K)	~12 (typical)

These data are obtained from Cody *et al.* (1965).

† Cody *et al.* (1965).

Into the chamber pass the vaporised chlorides of niobium and tin together with hydrogen gas. At the hot surface of the ribbon the chlorides are reduced by the hydrogen to form hydrogen chloride and niobium and tin. The latter react together on the hot surface of the tape to form the compound Nb_3Sn. The hydrogen chloride is swept out of the chamber by the excess parent vapours.

The vapour-plated ribbon is then cleaned in a mixture of nitric and hydrofluoric acids before passing through an argento-cyanide plating bath in which $0 \cdot 0025$ to $0 \cdot 005$ cm of high-purity silver is deposited.

The thickness of the Nb_3Sn deposit is adjusted to match the range of field strengths in which it is to operate. For fields up to 65 kA cm^{-1}, $0 \cdot 0006$ cm of Nb_3Sn is formed on each side of the stainless steel. For operation in fields of 65 to 100 kA cm^{-1} the thickness is $0 \cdot 0009$ cm and for operation above 100 kA cm^{-1} the thickness is $0 \cdot 0012$ cm. Thus, in high field strengths which reduce the critical current density, the increased thickness of Nb_3Sn allows roughly the same saturation current as the thin deposit gives in lower fields. The structure of a typical vapour-plated niobium–tin ribbon is shown in Figure 9.6.2.

2. *Diffused-layer tapes.* As an alternative to vapour deposition, a diffusion layer of Nb_3Sn can be formed between niobium and tin. This diffusion technique is more widely used than vapour deposition. Basically, a niobium ribbon, of width between $0 \cdot 63$ cm and $1 \cdot 27$ cm and of thickness $0 \cdot 0025$ cm is passed through a bath of molten tin which adheres to its surface. The tape is then passed through a reaction chamber containing an inert gas at low pressure in which the coated niobium is heated to 950°C for 15 min. During this treatment tin diffuses into the niobium to a depth of up to $0 \cdot 001$ cm forming the compound Nb_3Sn. High-conductivity copper ribbon is then soldered to one or both sides of the tape. Finally,

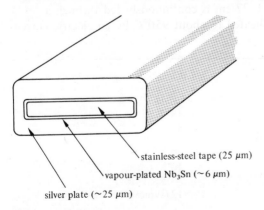

stainless-steel tape (25 μm)

vapour-plated Nb_3Sn ($\sim$6 μm)

silver plate ($\sim$25 μm)

Figure 9.6.2. The structure of a typical vapour-plated niobium–tin ribbon. The cross-section is not to scale and the width may be from $0 \cdot 25$ cm to $1 \cdot 27$ cm.

this copper-clad tape may be coated with a film of insulating varnish. The structure of a typical diffused-layer niobium–tin ribbon is shown in Figure 9.6.3, together with the curve of saturation current as a function of field strength. Plate III shows a few of the many types of niobium–tin ribbon now commercially available.

Vapour-plated and diffused-layer superconductive ribbons are similar in their operating characteristics; the vapour-plated ribbons have generally thicker Nb_3Sn layers and can be obtained in narrower widths than the diffused-layer tapes. The narrow width allows the winding of coils in layers, in which the winding arrangement is roughly as follows. Layers of ribbon are interleaved with a sandwich of Melinex (polyester film) and high-conductivity copper foil. This latter provides fairly good conductive cooling for the centre turns of each layer and also provides magnetic damping. In this function the foil serves the same purpose as the silver plate, even though it is not electrically coupled to the superconductor. The turns of each layer are shorted to a narrow strip of copper foil which lies parallel to the axis of the coil. These strips of copper foil limit the voltages that are generated during quenching. Electrically, these strips represent a distributed resistance which gives the coil a long response time. Thus fluctuations in the voltage across the coil produce currents in the strips and these currents can produce significant heating with consequent degradation in the performance of the coil. Hence, layer-wound coils are

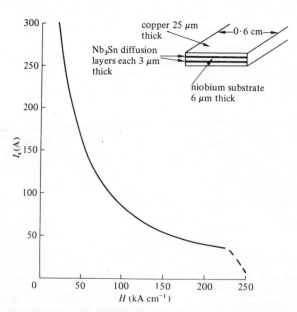

Figure 9.6.3. The saturation current plotted as a function of field strength at $4 \cdot 2°K$ for a typical diffusion layer niobium–tin superconductive ribbon. The dimensions and structure of the ribbon are shown in the inset.

intolerant of even small alternating voltages and rapid charging rates. Like disc-wound coils, to be described next, layer-wound coils do not usually operate at the saturation current.

Diffused-layer Nb_3Sn ribbons are made in widths of $0\cdot63$ and $1\cdot27$ cm. Superconducting magnets using the $0\cdot63$ cm wide material are composed of disc-wound coils, cooled generally at their edges. Face cooling is not easily achieved without sacrifice of filling factor, for ample space is needed between turns for the escape of bubbles of helium gas. Furthermore, a porous structure is not compatible with mechanical rigidity which is of considerable importance in view of the large forces exerted on the winding. Edge-cooled disc coils therefore constitute the majority of magnets using diffused-layer Nb_3Sn ribbon.

To illustrate the limitations on the overall current density of a disc coil imposed by the dynamic stability of the ribbon, we shall collect together the values of J_{cs} and $H_\perp$ predicted by the various criteria of Section 9.5. We shall consider the ribbon whose dimensions and curve of saturation current as a function of field are shown in Figure 9.6.3. The values assigned to the parameters of the ribbon are as given before in Section 9.5, with the addition of the following values:

$d_w = 15 \times 10^{-4}$ cm,
$J_c = 2\cdot5 \times 10^6$ A cm^{-2} at low field strengths,
$h = 1\cdot0$ W cm^{-2} degK^{-1},
$\tau = 10°$K,
$c = 4\cdot5 \times 10^{-3}$ J cm^{-3}.

1 Perfect cooling, J_{cs} limited by thermal conduction. Equation (9.5.10) gives, as before,

$\qquad J_{cs} = 6\cdot9 \times 10^5$ A cm^{-2}.

2 Perfect thermal conduction, finite heat transfer coefficient. Equation (9.5.19) gives, as before,

$\qquad J_{cs} = 1\cdot14 \times 10^5$ A cm^{-2}.

3 Perfect cooling, limiting perpendicular field. Equation (9.5.14) gives

$\qquad H_\perp = 29$ kA cm^{-1}.

4 Perfect thermal conduction, finite heat transfer coefficient. If we take J_c as $2\cdot5 \times 10^6$ A cm^{-2}, Equation (9.5.20) gives

$\qquad H_\perp = 0\cdot22$ kA cm^{-1}.

5 Maximum stably shielded field as given by Equation (9.2.14)

$\qquad H_\perp = 7\cdot8$ kA cm^{-1}.

6 Perfect thermal conduction, finite heat transfer coefficient as case 4 but $J_c = 2\cdot5 \times 10^5$ A cm^{-2} corresponding to a high parallel field strength.

Equation (9.5.20) gives

$$H_\perp = 2\cdot2 \text{ kA cm}^{-1} .$$

These six cases refer to edge-cooled ribbons only. It is clear that the limit of stability of the ribbon is dictated in this example by the heat transfer coefficient rather than by thermal conduction in the Nb_3Sn layer. This will in fact remain so within the range of the inequality

$$\tfrac{1}{4}\pi^2 K_n > hw .$$

For the particular ribbon this sets a value of about 25 cm to the half-width of the ribbon in which thermal conduction matches the effect of heat transfer coefficient. However, in practice, because of the uneven face of a disc coil, the surface area actually available for heat transfer is probably as much as ten times greater than the theoretical value. This would increase the values given in cases 2, 4, and 6.

Now we assumed that the actual value of J_c in the Nb_3Sn layer was $2\cdot5 \times 10^6$ A cm^{-2}. Hence case 2 shows that this actual critical current density exceeds the permitted stable value according to criterion (9.5.19) and the ribbon will be unstable. However, even at a current density of $1\cdot1 \times 10^5$ A cm^{-2} the perpendicular field strength which could just be reduced to zero within the half-width of the ribbon and in the absence of any instability would be given by Equation (9.5.15) as $7\cdot3$ kA cm^{-1}. This is greater than the maximum permitted perpendicular field given by 4 and 6; therefore one of these latter determines the limit of stability.

So, for the configuration chosen, the perpendicular field strength determines the limit of stability. If no parallel component of field strength is present to reduce the critical current density J_c, case 4 indicates a maximum permitted perpendicular field of $0\cdot22$ kA cm^{-1}. However, the inherent stabilising effect of the heat capacity of the ribbon gives a permitted perpendicular field of $7\cdot8$ kA cm^{-1} as shown by case 5.

It is found in practice that disc coils tightly wound with this type of ribbon will operate at overall current densities of about 20×10^4 A cm^{-2} and that perpendicular fields of up to 20 kA cm^{-1} can be tolerated. This is due of course to the effective increase in the area of the cooled surface arising from unevenness in the edge of the ribbon.

If increased overall current density is sought, face cooling must be employed, despite the mechanical disadvantages this introduces. The heat transfer coefficient limits the permissible critical current density in this case. From Equation (9.5.18) J_{cs} is given as $1\cdot67 \times 10^6$ A cm^{-2}. Because of the need for adequate cooling space between ribbons the overall maximum current density will be about 80×10^3 A cm^{-2}.

We may summarise the limits of stability of ribbon-wound disc coils as follows.

Very high critical current densities or large thicknesses of Nb_3Sn are permissible if the ribbon is effectively face cooled. The limiting values of the critical current density or, alternatively, of the thickness of the super-conductive layer may in this case be determined by either the thermal conductivity or the heat transfer coefficient. Typical values of stable critical current density and layer thickness under these conditions are respectively 10^6 A cm^{-2} and 3×10^{-4} cm.

Edge-cooled ribbons will by contrast be stable only at low critical current densities (10^5 A cm^{-2}) and low perpendicular field strengths (2 to 20 kA cm^{-1}). By virtue of its thermal capacity alone a typical ribbon will stably shield a perpendicular field of about 8 kA cm^{-1}.

The thermal conditions in a disc coil are rather easily definable; thus, as we have assumed, either face or edge cooling prevails. In layer-wound coils the conditions affecting stability are less easily definable. Liquid helium cannot readily penetrate between the turns of a layer-wound coil and the only cooling mechanism available is the conduction of heat from the face of the ribbon through a thin layer of insulating film (for instance, Melinex) to a copper foil. This gives essentially a transient heat transfer mechanism, for the copper foil heats up locally and the effective heat transfer coefficient falls. However, by considering the ribbon to be simply face-cooled, we may estimate the permissible critical current density.

We set d_n equal to the combined thickness of the normal metal layer on one side of the ribbon and half the thickness of the copper foil. Thus, for example, $d_n = 30 \times 10^{-4}$ cm. Next we calculate a heat transfer coefficient given by

$$h = \frac{K_m}{t_m} ,$$

where K_m and t_m are, respectively, the thermal conductivity and thickness of the insulating film of Melinex. For example, this is typically $0 \cdot 1$ W cm^{-2} degK^{-1}. The ribbon used in layer-wound coils is narrower and has a thicker layer of Nb$_3$Sn than that illustrated in Figure 9.6.3. We shall take d_s to be 6×10^{-4} cm. Substituting these values into Equation (9.5.18), we obtain $J_{cs} = 2 \cdot 8 \times 10^5$ A cm^{-2}. In fact the actual critical current density in low field strengths is about $2 \cdot 5 \times 10^6$ A cm^{-2}. Therefore we would expect the ribbon to be unstable. This is observed in practice. However, transport current densities of up to 3×10^5 A cm^{-2} can be carried in the Nb$_3$Sn layer.

It appears that, when the condition for dynamic stability is violated, complex transient processes occur with momentary current flow in the normal metal layer and transient heat transfer to the copper interleaving. The agreement between computer studies of such processes and experimental measurements indicates that the limiting transport current density in layer-wound coils and possibly also in disc coils is indeed determined in this way.

9.7 Superconducting magnets

Because of the very important and burgeoning application of superconductors to the construction of magnets three particular magnets will
be described briefly.

9.7.1 The bubble chamber magnet at the Argonne National Laboratory†

This magnet has been built for the $3 \cdot 7$ m liquid hydrogen bubble chamber
at the Argonne National Laboratory in Illinois. Its major parameters are
listed in Table 9.7.1; illustrations showing its location around the bubble
chamber and its general construction are given in Figure 9.7.1. and
Plate IV.

It is the biggest superconducting magnet in terms of weight and stored
energy yet built and for these characteristics alone deserves mention in
this book. It also illustrates well, however, some of the considerations in
the design of cryostatically stable magnets.

Although the total weight of the coils is 45000 kg, only 400 kg of this
is superconductor. The great preponderance of copper over niobium–
titanium is needed to give the full stabilisation and the strength to
withstand the magnetic forces.

The conductor takes the form of a composite strip, 5 cm wide and
$0 \cdot 25$ cm thick containing six strands of niobium–48 wt.% titanium, whose
combined saturation current at $2 \cdot 5$ T induction is 4000 A. The composite
is operated at 2200 A in a peak induction of $2 \cdot 0$ T at the winding.

The magnet consists of four sets of disc coils containing a total of
thirty discs. The axis of the coil is vertical. Each disc coil is composed
of a strip of composite conductor and an insulating strip of polytetra-
fluorethylene. An epoxide resin bonds the conductor to the insulator to
form a rigidly integrated coil. This form of construction permits only
edge cooling of the composite superconducting strips.

Because of the size of the magnet, the operating current must be well
below the recovery current. The heat transfer rate needed to maintain
stability at 2200 A is given by Equation (9.4.2) as $0 \cdot 13$ W cm^{-2}. This is
adequately less than the recovery heat flux of $0 \cdot 4$ W cm^{-2} for this
configuration.

Table 9.7.1. $1 \cdot 8$ T magnet parameters.

Field strength	$1 \cdot 8$ T
Coil inside diameter	$4 \cdot 8$ m
Coil outer diameter	$5 \cdot 3$ m
Coil height	$3 \cdot 0$ m
Operating current at $1 \cdot 8$ T	2200 A
Inductance	40 H
Energy stored	80 MJ
Weight of coils	45000 kg

† Laverick (1968).

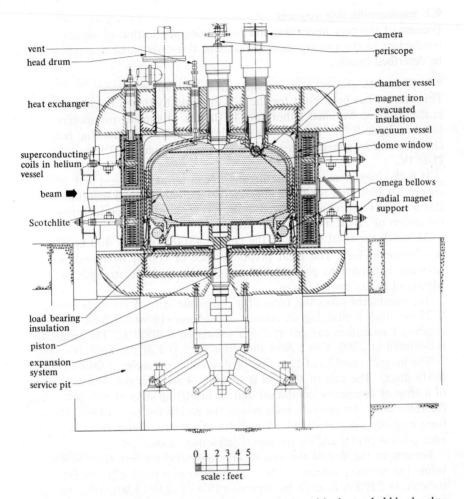

Figure 9.7.1. The general arrangement of the Argonne liquid hydrogen bubble chamber, with the 1·8 T superconducting magnet.

There is another criterion for the amount of copper in the composite. The hoop stress in the conductor is given by the simple expression

$$\sigma_H = BJR \ ,$$

where B is the magnetic induction at the conductor in teslas, J is the current density in the conductor in amperes per square metre and R is the radius of the conductor in metres. The stress will then be in newtons per square metre. For the Argonne magnet the stress is $0 \cdot 845 \times 10^8$ N m^{-2}. (For those who have not held a newton in their hands, it weighs $0 \cdot 225$ lb. The above stress is therefore $12\,200$ lb in^{-2}.) Such a stress is high for annealed copper. But it must be remembered that the strands of niobium–titanium are relatively much stronger than the copper and can support a considerable part of this hoop tension.

If we apply the stability condition (9.5.21) to the superconducting strands in the present composite conductor, we find that the condition for the commutation of the current from the copper to the superconductor following a flux jump is not satisfied. However, in order to apply Equation (9.5.21) we must assume that the strands are circular in cross section. In fact they are considerably flattened thus allowing heat generated within the strands to be discharged to the copper with less temperature differential than would be required if the strands were round.

Another criterion which this magnet violates without apparent ill effects is the temperature rise resulting from the collapse of gross shielding currents. In Section 9.4 we calculated this temperature rise to be 24 degK. As a result of such a temperature rise the resistivity of the copper would rise to about $2 \cdot 6 \times 10^{-8}$ Ω cm. This would increase the heat transfer rate to $0 \cdot 26$ W cm^{-2}. However, this is sufficiently below the recovery heat flux to allow a return to superconductive operation. Furthermore, a flux jump involving the simultaneous collapse of shielding currents in all six superconducting strands is most unlikely to occur. Because each of the strands of superconductor is quite large, the shielding currents flowing in it will collapse spontaneously at low current levels. In this way the temperature rise during any flux jump will be limited to a few degrees Kelvin.

Through the agency of radiative heat transfer, resistance of electrical connections, and thermal conduction along mechanical supports this magnet requires about 500 W of refrigeration at $4 \cdot 5 °$K.

9.7.2 High-homogeneity magnet for nuclear magnetic resonance

Among the applications of superconductivity to research in physics, the high-homogeneity magnet is especially interesting.

The technique of nuclear magnetic resonance requires field strengths in the range 1 to 10 T with a homogeneity of at least 1 part in 10^5 over a volume of a cubic centimetre. Conventional iron-cored electromagnets will achieve such a homogeneity up to 2 T but only with complicated and

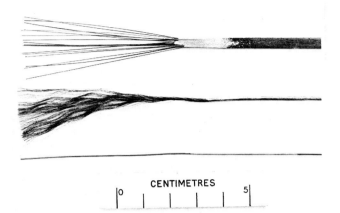

CENTIMETRES

Plate II. Some typical copper-clad niobium–titanium wires.
Left: a partially stabilised composite containing sixteen strands of superconductor and having a recovery current of 500 A at a field strength of 48 kA cm^{-1}.
Centre: an intrinsically stable wire containing 61 fine filaments of superconductor. This wire has an overall diameter of 0·04 cm and has a saturation current of 40 A at a field strength of 50 kA cm^{-1}.
Right: a single cored wire having a superconductive core of diameter 0·025 cm and an overall diameter of 0·04 cm.
The saturation current of this wire is typically 50 A at a field strength of 50 kA cm^{-1}.

Plate III. Some typical superconductive niobium–tin ribbons.
Left: a vapour-deposited ribbon of 0·25 cm width. Depending on the thickness of the superconducting layer the saturation current of this ribbon at a field strength of 80 kA cm^{-1} will vary between 50 and 250 A.
Centre: a diffused layer ribbon of 0·5 cm width having a typical saturation current at 80 kA cm^{-1} of 100 A.
Right: a diffused layer ribbon having a width of 1·27 cm and a typical saturation current at 80 kA cm^{-1} of 250 to 1000 A, depending on thickness.
This ribbon is considerably thicker than the others because of a layer of stainless steel applied to one side for strength.
Ribbons made by the General Electric Corporation.

Plate IV. A high homogeneity magnet wound from single core composite niobium–titanium wire. The two visible windings are compensating coils. These correct the errors in the central field of the main winding. The central field strength of this magnet is 60 kA cm^{-1} homogeneous to 2 parts in ten million over a sphere of 1 cm diameter. The thermal switches and superconducting joints are housed under the top cap.

Courtesy of the Oxford Instrument Company Limited

Plate V. A superconductive magnet using niobium–tin ribbon. A central field strength of 80 kA cm^{-1} is generated in a bore of 4 cm. It will be noticed that the end discs are spaced more widely than the centre discs. This lowers the radial field strength and improves the cooling, both of which effects help to stabilise the ribbon. At the bottom can be seen the housing rectifier flux-pump which generates 250 A to energise the magnet.

Courtesy of the General Electric Corporation.

Plate VI. The Argonne bubble chamber magnet in course of construction. In the foreground the lower half of the coil is being prepared for lowering into its stainless steel vacuum tank (shown covered in the middle background). In the far background the upper coil pair is seen welded into its vacuum tank.
Courtesy of the Argonne National Laboratory.

expensive power supplies. One characteristic of superconducting magnets enables them to achieve this homogeneity at fields well above 2 T, virtually without any power supply at all. The characteristic referred to is the ability of a superconducting magnet to operate in 'persistent mode'. This, as its name implies, is the persistent flow of a superconducting current in a wholly superconducting circuit.

Figure 9.7.2 shows the circuit of a persistent mode magnet. The crux of this circuit is the thermal switch. This is a length of superconducting wire wound non-inductively in an enclosure which can be heated to about 20°K. The superconducting wire is connected across the ends of the magnet winding through superconductive joints.

When the thermal switch is heated above 10°K the superconducting wire, either niobium–titanium or niobium–tin, is driven normal. In this condition it allows charging current to flow into the magnet winding. The resistance of the switch in its 'open' condition should be as high as possible to avoid appreciable energy loss due to shunt current flowing in it.

When the required operating current has been reached in the magnet the switch heater is de-energised and the switch allowed to recover its super-conductivity. If the external energising current is now slowly reduced to zero, the magnet current will flow through the switch without loss. At least, almost without loss.

It will be recalled that current flow in a hard, type II superconductor is never completely without loss because of flux creep. However the losses in the magnet and hence the decay of field will cease to be significant after a few minutes.

Plate V shows a typical high-homogeneity magnet for experiments on nuclear magnetic resonance. It generates a field of 7 T with a homogeneity of 1 part in 10^6 over a sphere of 1 cm diameter. The conductor used in this magnet is a co-drawn composite of niobium–titanium and copper.

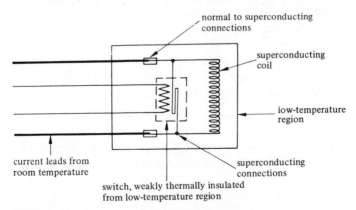

Figure 9.7.2. Circuit of a persistent mode magnet.

The diameter of the core is 0·25 mm and the overall diameter on the copper cladding is 0·40 mm. Such a conductor quenches typically at 30 A in the environment of a coil: this is less than its short sample current, indicating that the conductor is not dynamically stable. This can be readily and unequivocally confirmed by inserting into Equation (9.4.11) the characteristics of the conductor in low field strengths.

This is typical of small superconducting magnets wound with a single-cored niobium–titanium–copper composite.

Magnets of similar sizes and applications are now being wound from intrinsically stable niobium–titanium wire.

9.7.3 Niobium–tin magnets

Because of the relatively greater cost of niobium–tin superconductors, this material is not at present used in magnets of large volume. Its main application lies in magnets generating more than 72 kA cm^{-1} for which niobium–titanium is not suitable. Niobium–tin ribbon can be layer wound, like small-diameter niobium–titanium wire, or disc wound.

For layer winding, a 2·5 mm wide ribbon is generally used. The thickness of the ribbon is about 10^{-2} cm and between layers a sandwich of Melinex and copper foil is used to increase magnetic damping and to provide a thermal sink for heat generated by flux jumping in the super-conductor. This form of winding is quite widely used although slowly being superseded by the disc winding.

A typical disc-wound niobium–tin coil is shown in Plate VI. It generates a field of 88 kA cm^{-1} in a bore of 4 cm at a current of 200 A. The material is 1·27 cm wide and is similar in its structure to that shown in Figure 9.6.3. Because the strip is wound tightly into disc coils with a varnish insulation between turns, only edge cooling is available to the coils. As was seen in Section 9.5 such a coil is expected to be unstable in the presence of sufficiently large radial components of magnetic field strength. In order to reduce the radial field strengths to levels below the stably shielded field, the end disc coils of the magnet are spaced apart.

The magnet is energised by a flux pump of the type described in Section 10.6. The flux pump is located beneath the magnet and sufficiently far from it that the stray field of the magnet does not affect the operation of the power cryotrons in the flux pump. The primary current to the flux pump at full field is 5 A.

9.7.4 The superconductive homopolar motor†

It may have been correctly interpreted that most contemporary applications of superconducting magnets lie in the area of physics research. There is, however, one notable industrial application and that is to homopolar motors and generators.

---

† Appleton and Ross (1969).

The homopolar motor is a very simple device in which a conducting disc rotates between the poles of a magnet. The Lorentz force is provided by a current flowing radially through the disc and through the field of the magnet. Unlike the heteropolar motor, current collection is through slip-rings and the construction is simple and rugged. If the motor is correctly designed, there is no mechanical reaction between the rotor and the field. However, the homopolar motor suffers from the disadvantage of low voltage, high current operation.

Until recently these disadvantages prevented its use in most commercial applications, but the development of high-field superconducting magnets and of a segmented rotor have restored its competitive position.

To test the commercial viability of the superconductive homopolar motor, a unit rated at 3250 hp has been installed in the 2000 MW power station at Fawley, Hampshire to circulate feed water to the boilers. The superconducting magnet for this motor has the parameters shown in Table 9.7.4.

The homopolar motor is a particularly suitable application for a superconducting magnet because of the need for the highest possible field strength in a large volume and because of the absence of mechanical reaction between the field and the rotor. The most severe cryogenic problem in the construction of large superconducting magnets is associated with the forces between the magnet at low temperature and its surroundings at room temperature. No such forces occur in the homopolar motor.

It is to be expected that the widest industrial application of super-conducting magnets in the near future will be in this area.

Table 9.7.4. Homopolar motor magnet parameters.

Field strength	3·5	T	Operating current at 3·5 T	725	A
Coil inside diameter	240	cm	Inductance	55	H
Coil outer diameter	284	cm	Energy stored	1·46	MJ
Coil length	53·5	cm	Weight of coil	5500	kg

Superconductive a.c. power devices

10.1 Power devices†

In the search for applications of superconductors it is inevitable that the power loss in resistive elements of a.c. systems should have been given early consideration. We have seen how the application of the property of zero resistance can raise the efficiency of solenoids from 0 to 100%. It would have been tempting to apply superconductors to the reduction of the power loss in the generation and transmission of a.c. power, a mean loss for instance of about 500 MW in Great Britain.

However, it was found early in the study of the potential a.c. applications of superconductors that, far from being reduced, the inefficiency of a power system was likely to be increased by the use of superconductors. Why this should be is explained in the following section: it would promptly end the examination of the a.c. power applications of superconductors, were it not for two other important properties. These are the very high current densities which can flow in superconductors and the very high field strengths in which type II superconductors will operate.

These two characteristics of superconductors promise a substantial reduction in the size and weight of some a.c. power devices, which can be of great importance in air-borne and space systems. With mainly these applications in view, development work is proceeding on superconducting alternators. The use of superconductors in land-based power transformers is also receiving attention in the hope that a decrease in size and weight, and hence in cost, may be obtained; the reason for the pursuit of super-conductive power transmission lines is the need for a cable of low loss and low capacitance.

We shall first consider the effect of operating superconductors in a.c. conditions and then examine the applicability of superconductors to three types of power device.

10.2 a.c. operation of superconductors‡

Superconductors can carry a current either as a sheath current as in type I below H_c (or in type II below H_{c1}) or as a bulk current in hard, type II materials. The sheath current in a type I superconductor is limited by the bulk critical field H_c. The critical sheath current in a round wire in a magnetic field is given by

$$I_{cs} = 2\pi R(H_c - 2H_e),$$ (10.2.1)

where H_c is the bulk critical field of the superconductor in A cm^{-1}, H_e is the ambient external field strength perpendicular to the wire, and R is the radius of the wire in centimetres. Equation (10.2.1) also applies to

---

† Comm. I, Intern. Inst. Refrig. (1969).   ‡ Hancox (1966b).

type II superconductors if H_{c1} is substituted for H_c. If the external field is zero,

$$I_{cs} = 2\pi R H_c .$$ (10.2.2)

This is the Silsbee current referred to in Section 2.1.

For niobium, H_{c1} is roughly 1200 A cm^{-1}, so that I_{cs} cannot be greater than 7500 A cm^{-1} of radius. Furthermore, the overall current density decreases as the wire diameter increases. Low overall current density together with low critical field renders the type I superconductor unattractive for most power applications even though it can carry the sheath current without loss at frequencies up to a few thousand megahertz (see Section 5.4).

Hard, type II superconductors, by contrast, can carry large bulk currents at high overall current densities and in very high fields. But in a.c. conditions an appreciable power loss is involved.

The prediction of the power loss in a hard, type II superconductor involves a fairly simple although somewhat inexact procedure. The energy loss resulting from one magnetic field cycle is given by the area of the magnetisation loop of the superconductor as for instance in Figure 8.3.2. The power loss follows directly as the loop area.

The approximate loss per cycle can be calculated quite simply using the concept of the critical state. Consider Figure 10.2.1 which shows a stage in the penetration of an external field into a hard superconducting plate. We shall suppose that the plate is thin and that in consequence the field strength does not vary greatly in the plate. Then we can assume that the current density is constant for any one value of external field strength.

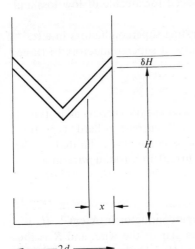

Figure 10.2.1. The penetration of an a.c. magnetic flux into a hard superconducting plate (see text).

The calculation proceeds in a somewhat similar manner to the calculation of magnetothermal stability. We assume that there has been a small increase dH in the external field strength: then at a distance x from the midplane of the plate a quantity of flux $d\phi$ crosses a line of unit length such that

$$d\phi = \mu_0 J_c x \, dH . \tag{10.2.3}$$

This flux movement generates a voltage impulse

$$V dt = d\phi = \mu_0 x \, dH . \tag{10.2.4}$$

This impulse interacts with the local current density to generate a density of heating $Q(x)$ given by

$$Q(x) = \mu_0 x \, dH . \tag{10.2.5}$$

Thus the total heat dissipation per unit length of plate and over the whole thickness and width of the plate is

$$Q = 2 \int_0^{d_s} 2wQ(x)dx = 2\mu_0 J_c d_s^2 w \, dH , \tag{10.2.6}$$

where $2w$ is the width of the plate.

Now, although J_c is assumed constant through the plate, it will vary with external field strength. Thus we use the critical-state equation

$$J_c \times (B + B_0) = \alpha_c . \tag{10.2.7}$$

Substituting for J_c in Equation (10.2.7), we get

$$Q = 2\mu_0 w d_s^2 \alpha_c \, dH (B + B_0)^{-1} . \tag{10.2.8}$$

The heat dissipated between zero field and a maximum field H_m is

$$Q = 2w d_s^2 \alpha_c \ln\left(\frac{\mu_0 H_m + B_0}{B_0}\right) . \tag{10.2.9}$$

Now a complete field cycle consists of four similar periods of dissipation, as can be seen from the hysteresis loop of Figure 8.3.2. Therefore the total loss per cycle is $4Q$ and the loss rate is of course

$$W = 8w d_s^2 \alpha_c f \ln\left(\frac{\mu_0 H_m + B_0}{B_0}\right) . \tag{10.2.10}$$

f is the frequency in hertz. The most significant term in the equation is the square of the thickness. The need for very thin sections of superconductor is clearly implied.

The parameters which may be adjusted so that the loss is minimised become clear if the equation is recast in terms of current. Thus,

$$W = 2I_c d_s f (\mu_0 H_m + B_0) \ln\left(\frac{\mu_0 H_m + B_0}{B_0}\right) , \tag{10.2.11}$$

where I_c is the critical current of the superconductor at the field H_m. In the expression I_c and d_s may be considered independent parameters. To minimise the loss rate these should all be as small as possible. However, because the superconducting ribbon must carry a specific transport current, if it is to be of practical use, the critical current must be greater than the peak value of this transport current. Indeed, because of the decrease in pinning strength caused by the rise in temperature due to the dissipation, the critical current should be appreciably greater than the peak transport current.

For a given value of critical current the thickness may be reduced if either the pinning strength α_c or the width w is increased. Therefore, for minimum losses, these latter should be as large as possible and the thickness correspondingly as small as possible. Thus, a thin flat strip incurs less a.c. loss than a round wire, provided the field direction is maintained accurately parallel to the wide face of the ribbon.

If an alternating transport current flows, in phase with the alternating field, the expression for the loss rate is modified slightly. The effect of the transport current is to create asymmetry in the fields on either side of the strip. We assume that the difference is small in comparison with the peak field strength H_m but that the major effect of the transport current is to shift the point of minimum flux density within the ribbon.

Then the expression for the total heat dissipation per unit length of the plate and over its whole width and thickness becomes

$$Q = \int_0^{d_s(1 \pm J_t/J_c)} \mu_0 J_c x \, dx \, dH \ , \tag{10.2.12}$$

where $J_t = I_t/4d_s w$, I_t being the instantaneous transport current.

If the peak transport current is denoted I_m, then

$$\frac{I_t}{I_m} = \frac{H}{H_m} \ .$$

Integrating, we find

$$Q = \mu_0 J_c \, dH d_s^2 w \left(1 \pm \frac{J_t}{J_c}\right)^2 \ . \tag{10.2.13}$$

Substituting for J_t as

$$J_t = I_m H (4d_s w H_m)^{-1} \tag{10.2.14}$$

and for J_c as before and integrating with respect to H, we finally obtain for the total loss per cycle

$$Q = 2\alpha_c w d_s^2 \left\{ \ln\left(\frac{\mu_0 H_m + B_0}{B_0}\right) + \frac{I_m^2}{I_c^2}\left[\frac{\mu_0 H_m (3\mu_0 H_m + 4B_0)}{12(\mu_0 H_m + B_0)^2}\right] \right\} \ . \tag{10.2.15}$$

From Equation (10.2.15) it can be seen that, for all values of B_0 and $\mu_0 H_m$, the presence of a transport current in phase with the external a.c. field increases the losses only slightly. If, as would probably be the case, the peak a.c. transport current is much less than the critical current, the loss is only marginally greater than that for zero transport current.

For an alternating transport current in 180° phase opposition to the field, the expression for the loss rate is identical with that derived immediately above. Transport currents of other phase angles give rise to augmented losses that are slightly greater than for an in-phase current. The augmented hysteresis loss is a maximum for a current leading or lagging by 90°. Very approximately, transport current augments the hysteresis loss by a factor $(I_m/I_c)^2/4$. However, it is clearly necessary that I_m is appreciably less than I_c for reasons already mentioned, so that for all practical purposes the effect of transport current upon the losses can be ignored.

By way of example we shall calculate the hysteresis loss per centimetre length of the ribbon of Figure 9.6.3 at 50 Hz, ignoring the effect of the copper cladding. We shall assume that the ribbon experiences an alternating field strength of peak amplitude 40 kA cm^{-1} oriented exactly parallel to the wide face of the ribbon. We shall ignore the effect of transport current. The critical current at 40 kA cm^{-1} is 200 A and B_0 is $1 \cdot 6$ T. Then from Equation (10.2.11) the hysteresis loss rate is given by

$$W = 5 \cdot 6 \text{ mW cm}^{-1} .$$

Suppose now, however, that the field is not exactly parallel to the ribbon but inclined at an angle to the face. The field component perpendicular to the face will incur a hysteresis loss which may be calculated quite simply. The loss is now given by

$$W' = 8w^2 d_s \alpha_c f \tan \psi \ln \left(\frac{\mu_0 H_m + B_0}{B_0} \right) , \qquad (10.2.16)$$

where ψ is the small angle between the face of the ribbon and the field. H_m is the peak field strength. If this component of the loss is to be no greater than that due to the parallel field component, then

$$\tan \psi = \frac{d_s}{w} .$$

In the case of the ribbon of Figure 9.6.3, $d_s/w = 5 \times 10^{-4}$. Thus the perpendicular component of field must not be greater than 20 A cm^{-1} if the hysteresis loss is not to be more than doubled.

These instances of a.c. hysteresis loss in hard superconductors in large alternating field strengths are such as might be found in the armature of a superconducting alternator, for example. We shall use the expressions to calculate the probable efficiency of a wholly superconducting alternator in a later section.

Another case of particular interest is an alternating current in a flat strip, with negligible incident field and no edge effects. Such a configuration might be found in a superconducting power cable. In this case we consider the critical current density to be constant at all times. This approximation is justified by the small variation in field strength produced by the transport current. Again, consider the strip shown in Figure 10.2.1. The magnetic flux crossing unit length of the strip at a distance x from the point of zero field strength is given by

$$\phi(x) = \mu_0 x \, dH .$$

where dH is a small increase in external field strength. This flux movement couples with the current density, which is of course the critical value J_c, to cause a local energy dissipation

$$Q(x) = \mu_0 J_c x \, dH . \qquad (10.2.17)$$

The total energy dissipated in the distance of penetration δ, in the full width $2w$ and per unit length of strip, is

$$Q(\delta) = 2 \int_0^\delta Q(x) \, dx = 2 J_c \delta^2 \mu_0 w \, dH . \qquad (10.2.18)$$

Now δ is given by

$$J_c \delta = H ,$$

where H is the instantaneous external field strength. Hence the total energy dissipation per unit length of strip for a rise of external field from zero to H_m is given by

$$Q(H_m) = 2\mu_0 w J_c \int_0^{H_m} \frac{H^2}{J_c^2} \, dH = \frac{2\mu_0 w H_m^3}{3 J_c} . \qquad (10.2.19)$$

Now we must substitute for H_m in terms of the peak transport current I_m. Thus,

$$H_m = \frac{I_m}{4w} . \qquad (10.2.20)$$

Hence

$$Q(H_m) = \frac{\mu_0 I_m^3}{96 J_c w^2} . \qquad (10.2.21)$$

Substituting for J_c in terms of I_c, the critical current, we get

$$Q(H_m) = \frac{\mu_0 d_c I_c^2 q^3}{24w} , \qquad (10.2.22)$$

where

$$q = \frac{I_m}{I_c} .$$

The loss rate per metre follows by multiplying by the frequency:

$$W = \frac{\mu_0 d_s I_c^2 q^3 f}{24w} .$$

(10.2.23)

As before this hysteresis loss can be reduced by increasing the pinning strength (α_c) and thereby reducing d_s or by increasing the width. As an example we shall consider the ribbon of Figure 9.5.1 assuming a critical current of about 500 A in zero field. At a frequency of 50 Hz an alternating current of peak amplitude 250 A dissipates 72 μW cm^{-1} in hysteresis losses.

This loss rate is 1/100 of that for an alternating field of 40 kA cm^{-1} incident on the same ribbon. The difference is easily understood qualitatively if it is noted that a current of 250 A generates a field of only 200 A cm^{-1} at each surface of the ribbon. This shows also that the assumption of constant critical current density is totally justified in this case. It must be remembered, however, that in all the calculations involving flux penetration into hard, type II superconductors the effects of surface currents at fields below H_{c1} have been ignored. In the present instance the losses would probably be reduced by this effect.

Although we have calculated the hysteresis losses for only two cases of the great many possible, these two represent typical approximations for hysteresis loss in flat ribbons.

It has been shown that the most important factor is the field strength rather than the transport current and that the angle the field makes with the face of the ribbon has a dramatic effect on the losses. In general it is sufficient to consider the hysteresis losses as being due either to an externally applied alternating field, disregarding transport current of no matter what phase relationship, or solely to an alternating transport current.

We have so far considered the power dissipation in an infinite plate and exemplified these losses by reference to a specific niobium–tin ribbon. The use of a niobium–titanium wire of diameter 0·025 cm at 50 Hz would give rise to a very large hysteresis loss, through the influence of the diameter which enters linearly into Equations (10.2.11) and (10.2.23). However, the intrinsically stable composite superconductor consisting of fine superconducting filaments in a normal metal matrix is potentially capable of operating under a.c. conditions with relatively small dissipation, at least at power frequencies. The intrinsically stable composite super-conductor is realised, as was mentioned in Section 9.6.2, as a bundle of niobium–titanium filaments embedded in a matrix of copper or cupro-nickel. It is the latter matrix material which is of interest for a.c. applications.

Just as it influenced the stability of the filamentary superconductor, the pitch of the twist of the filaments also strongly affects the a.c. power loss. If the pitch is long, the wire will behave as if the superconducting filaments were smeared out over the cross section of the wire. In that case the a.c. power loss would be the same as in a solid superconducting wire of appropriate pinning strength and of the same diameter as the overall diameter of the intrinsically stable wire.

If the pitch of the twist is short, however, the higher effective cross resistance of the matrix will limit the shielding currents to low levels. The magnetic flux will then be able to penetrate into the matrix of the fine wire without having to cut through the filaments. However, the flux will still penetrate the superconducting filaments themselves but will do so as if each filament were situated in a field increasing steadily on both sides. The dissipation in the filaments will then be much less than it would have been if the flux in the matrix of the wire had cut right through the filaments.

The power loss in the wire is thus seen to have two components: (1) the penetration of magnetic flux into the matrix of the wire; (2) the penetration of flux into the filaments. The eddy current loss in the matrix can be calculated from the expression

$$W_m = \tfrac{1}{8}\pi^3\mu_0^2 H_m^2 f^2 a^4 \rho^{-1} , \qquad\qquad (10.2.24)$$

where a is the radius of the wire, f is the frequency, and H_m is the peak field strength. (This formula is found in any standard textbook on electro-magnetic field theory.)

The hysteresis loss in the superconducting filaments is given approximately by Equation (10.2.11) with d_s set equal to the radius of the individual filaments. I_c is the total critical current of all the filaments at the peak field strength.

By way of example, to enumerate the typical magnitudes of these losses, we may consider a composite superconductor of overall diameter 0·025 cm containing a large number of superconducting filaments. We shall assume that the peak field strength is 40 kA cm^{-1}, that the critical current at this field strength is 100 A, and that the resistivity of the matrix of the composite is 3×10^{-5} Ω cm. (This corresponds to an alloy of cupro-nickel at very low temperatures.) The diameter of each filament is assumed to be 10^{-3} cm and the operating frequency is 50 Hz.

Firstly, to satisfy the condition that negligible shielding current should flow in the superconducting filaments the maximum pitch of the twist is given by Equation (9.4.14) as

$$l_c = 0\cdot87 \text{ cm.}$$

We must assume therefore that the wire has a twist with a pitch less than this value.

The eddy current dissipation in the matrix is given by Equation (10.2.24) as

$$W_m = 0\cdot2 \text{ mW cm}^{-1} .$$

The hysteresis loss is given very approximately by Equation (10.2.11):

$$W_h = 5 \cdot 4 \, \text{mW cm}^{-1} .$$

Thus it is seen that the hysteresis loss in the superconductor is the predominant mechanism of dissipation, at least in the hypothetical wire. It is comparable with the value calculated for the niobium–tin ribbon; this follows from the equality of the diameter of the filaments and the thickness of the ribbon and the similar values of critical current. However, the power loss in the filamentary superconductor is not sensitive to the orientation of the field as is that of a ribbon.

Composite superconductors containing niobium–titanium filaments, 10^{-3} cm in diameter, in a cupro-nickel matrix are already being manufactured and there appears to be no obstacle to the production of filaments of 10^{-4} cm in diameter. Consequently the hysteresis loss noted above could well be reduced by a factor of 10. Composites of niobium–titanium in cupro-nickel hold the promise of being viable components of superconductive a.c. power systems.

10.3 Superconductive rotating a.c. machines
Both superconducting alternators and motors have been constructed. Rather more attention has been paid to the development of alternators than to motors, as such, although it will be appreciated of course that alternators and synchronous motors are identical in their theoretical treatment if not in the details of their construction.

We shall begin by considering the particular problems attending the use of superconductors in alternators.

10.3.1 Superconductive alternators
It is mainly to this class of a.c. power device that attempts have so far been made to apply hard superconductors. Both the armature and rotor of an alternator are candidates for the application of superconductors although the problems involved are somewhat different.

The rotor of an alternator must supply the steady excitation field, constant in time but rotating within the armature. This is essentially a d.c. field and superconductors could be used for its generation, with almost no problems of hysteresis loss. There are small ripple components of field superimposed on the rotor field by the armature, but they are small and not likely to cause significant a.c. losses.

By contrast, the armature carries the full alternating currents that the machine generates and the hysteresis loss in this is of fundamental significance.

The size and weight of an alternator are determined by the optimum combination of its magnetic, electrical, and mechanical components. It is worth, at the outset, stating the basic functions of an alternator (or synchronous motor) with reference to these three components. The functions are as follows.

1 The production of torque, given by

$$\pi T = PNI\phi ,\qquad (10.3.1)$$

where NI is the armature current–turns product per phase, ϕ is the excitation flux of the rotor per phase, and P is the number of phases.

2 The production of back e.m.f. in the armature given per phase by

$$E = 2Nf\phi ,\qquad (10.3.2)$$

where N and ϕ are as before, and f is the frequency.

3 The production of power, whose mean value is given by

$$W = PEI = 2PNIf\phi .\qquad (10.3.3)$$

The parameters in this last expression determine the size and weight of a machine. Usually P, I, and f are specified leaving N and ϕ to be selected to give a machine optimised for minimum cost, volume, or weight. For minimum volume or weight N and ϕ must be large.

In conventional rotating machinery iron is introduced into the magnetic circuit to minimise volume and power loss caused by magnetising currents. Because of this iron, the magnetic induction B is limited to about $1 \cdot 5$ T and the desired value of ϕ must be achieved by using an armature of sufficient length and diameter. For a given rotational speed the diameter is limited by the centrifugal force on the rotor poles or conductors. However, the e.m.f. is given by $E \propto \omega D$, where ω is the rotational speed of the rotor and D the diameter of the armature, but the centrifugal force is given by $F \propto \omega^2 D$; therefore, limitation on force can always be met for a given e.m.f. by increasing D and decreasing ω. N is limited by the diameter of the armature and the permissible current density in the conductors.

There are therefore three ways in which superconductors may reduce the size and weight of alternators: (1) by increasing the excitation induction B; (2) by increasing the permissible current density in the armature winding; (3) by reducing the centrifugal force on the rotor conductors.

Superconductors can in theory at least achieve all these.

The rotor of an alternator is an immediate candidate for the use of superconductors. It has only very small high-frequency a.c. components, these being induced by the reaction of the armature currents. Because of the essentially steady magnetic conditions, superconductors will operate reliably in the rotor winding. Iron may then be dispensed with, for fields of 40 kA cm^{-1} can easily be achieved without it by superconductive windings of modest size and weight.

A small 4 kVA alternator using such a hybrid system of conventional armature and superconducting rotor has been built and tested (Stekly and Woodson, 1964). Although the armature surrounded the excitation field as in conventional alternators, the excitation field winding was stationary and the armature rotated around it. This machine was small and light but the Dewar housing the exciting field was considerably bigger than the machine itself. It demonstrated, however, that superconducting 'rotors' can be operated reliably in the presence of armature reaction fields with hardly any power loss. By contrast the use of superconductors in an armature winding introduces significant a.c. power losses. The magnitude of these losses in relation to the output of an alternator is easily estimated.

From Equation (10.2.11) the power loss in a metre length of superconductor of thickness $2d_s$ metres is given by

$$W_L = 2I_c f d_s (\mu_0 H_m + B_0) \ln \left(\frac{\mu_0 H_m + B_0}{B_0} \right) ,$$

where I_c is in amperes and H_m is in amperes per metre. If the peak operating current I is related to I_c by the relationship

$$KI = I_c ,$$

then

$$W_L = 2KIfd_s(\mu_0 H_m + B_0) \ln \left(\frac{\mu_0 H_m + B_0}{B_0} \right) .$$

Now in a superconductive alternator H_m might be about 40 kA cm^{-1} and for a Nb$_3$Sn ribbon B_0 is, as we have already seen, about 1·6 T. Therefore,

$$\ln \left(\frac{\mu_0 H_m + B_0}{B_0} \right) \approx 1$$

and, very approximately,

$$\mu_0 H_m + B_0 = \mu_0 H_m .$$

With these free substitutions the hysteresis loss is written

$$W_L = 2\mu_0 KIfd_s H_m . \tag{10.3.4}$$

The output power of the alternator, given by Equation (10.3.3), can be rewritten by substituting for ϕ in terms of armature diameter and field strength; then, per phase,

$$W_0 = 2\pi\mu_0 NIH_m lDf , \tag{10.3.5}$$

where H_m is the rotor field strength experienced by each armature conductor as the rotor field sweeps past it.

$2Nl$ is the total length of conductor per phase. Thus the power output per unit length of conductor per phase is

$$W_0 = \pi\mu_0 IH_m Df . \tag{10.3.6}$$

The hysteresis loss expressed as a fraction of the output is then given by

$$\frac{W_L}{W_0} = \frac{2}{\pi} K \frac{d_s}{D} .$$ (10.3.7)

This expression shows clearly that, if the electrical efficiency of the alternator is to be fairly high, the thickness of the superconductor must be very small compared with the diameter of the armature. It should be borne in mind also that the thickness d_s referred to must be measured at right angles to the direction of the field. Also, the hysteresis loss in the superconductor is incurred at low temperature. It must be rejected, however, at a high temperature. On the assumption, for the sake of example, that one watt of heat at $4 \cdot 2°$K requires 2000 W for its rejection at room temperature, then an alternator–refrigerator unit with an overall efficiency of 97% must have a ratio d_s/D of not more than $1 : 60000$.

It is instructive to calculate by way of example the approximate size and efficiency of a 1 MVA, 50 Hz superconductor alternator. We shall assume arbitrarily a three-phase output of 300 V and 1000 A per phase and that the armature is wound with the Nb$_3$Sn ribbon described in Section 9.6. The rotor field is assumed to be 40 kA cm^{-1}, parallel at all points to the plane of the ribbon, which is illustrated in Figure 9.6.3. From Equation (10.3.2) the r.m.s. voltage per phase can be deduced to be

$$V = 4 \cdot 44\pi f N B R^2 ,$$ (10.3.8)

where N is the number of armature turns per phase, R is the radius of the armature in metres, B is the rotor flux density in tesla, and in which it has been somewhat arbitrarily assumed that the active length of the armature is twice the radius. [From Equation (10.3.7) it would be concluded that R should be as large as possible: however, centrifugal forces on the rotor windings impose a limit to R.]

There will be altogether $3N$ turns ($6N$ conductors) spaced around the armature. Each conductor must carry 1000 A in a field of 40 kA cm^{-1} using ten ribbons per conductor so that each carries 100 A peak; the peripheral pitch of the armature conductor will be 1 cm, so that the minimum radius of the armature is given by

$$2\pi R = \frac{6N}{100} ,$$ (10.3.9)

R being in metres. Substituting this into Equation (10.3.8) and equating V to 330 V yields

$$330 = \frac{4 \cdot 44}{0 \cdot 03} \pi^2 f B R^3 ,$$ (10.3.10)

whence $R = 9 \cdot 7$ cm. This is very much smaller than the radius of a conventional 1 MW machine. The centripetal acceleration at 50 Hz is then 89 g which is an acceptable value.

To derive the hysteresis loss we note that the critical current of each of the ten ribbons in a conductor is 200 A. Thus $K = 0.5$, whence

$$W_L = \frac{2}{\pi} W_0 K \frac{d_s}{D} = 10.7 \text{ W}.$$

If we allow for a refrigeration ratio of 2000, the overall efficiency of the alternator would be 98%.

However, lest the dazzling prospect of a theoretical 1 MVA alternator, 18 cm in diameter and 18 cm long, blind us to reality, it will be mentioned again that the problem of maintaining the field parallel to the face of the ribbon is in itself formidable, to say nothing of the cryogenic problems associated with the use of moving superconductive elements. Also we have considered only the hysteresis losses in the armature. There would be many other sources of inefficiency. Generally, imperfection in the cryogenic system, electrical leads, and mechanical links to room-temperature systems would also introduce losses. The overall efficiency of the alternator would be unlikely to exceed 95%.

So, although the alternator itself is very much smaller and lighter than a conventional 1 MVA machine, it would require a 50 W refrigerator for continuous operation. Such a refrigerator would, to quite a large extent, detract from increased compactness of the superconducting alternator.

Despite the formidable cryogenic problems involved, the development of superconducting alternators is proceeding steadily. A small hybrid unit with superconducting field windings and conventional armature has already been mentioned.

A more ambitious project is a wholly superconducting system built as a prototype for a 1 MVA alternator (Dynatech Corporation, 1964). In this 50 kW was generated in a three-phase two-pole armature. The operating temperature was $10°$K. This reduces the critical current in the Nb_3Sn ribbon with which the armature and field coils are wound but the refrigeration requirements are also somewhat reduced. The total heat input to the $10°$K region was about 60 W, which would correspond to about 30 kW of refrigeration power. The overall efficiency of the system is therefore quite low but it is to be expected that in a larger system the net efficiency would be much greater than 40%.

10.3.2 Superconductive motors[†]

The foregoing examples of alternator design apply directly to synchronous motors. However, no units as large as 50 kW have been designed specifically as motors and most of the applications of superconducting motors have so far been of very low power.

[†] Schock (1961).

A very simple form of superconducting motor is the 'synchronous induction' motor. The principle of operation of the conventional induction motor is that currents induced in a rotor winding by the rotating field of the armature interact with this field to generate a driving torque. The rotor currents are proportional to the 'slip', that is the difference in rotational speeds of the armature field and of the rotor. Further, the 'slip' of the rotor is proportional to rotor resistance. If the rotor resistance is zero, the slip will be zero and the rotor speed will synchronise with the armature field rotation.

Figure 10.3.1 shows diagrammatically a simple superconducting motor of the 'synchronous induction' type. It consists of three superconducting field coils spaced at 120° around a closed superconducting loop which forms the rotor. A low-frequency three-phase supply to the field coils produces a rotating field. The rotor loop aligns itself in this field so that the current induced in it interacts with the field to provide the demanded torque, which is proportional to the lag angle α.

10.3.3 Operating characteristics of superconductive alternators
The behaviour of an alternator under normal and fault conditions is largely determined by its armature resistance and synchronous reactance. This latter is a combination of the self-inductance of the armature and the effect of cross magnetisation of the armature due to the current flowing in it. Machines with high synchronous reactance operate well in parallel and are most tolerant of fault conditions.

A superconducting alternator with high excitation field strength and no iron will have a low value of synchronous reactance. Consequently, fault currents will tend to be large. In order that the superconducting armature winding should remain superconducting during fault conditions, the

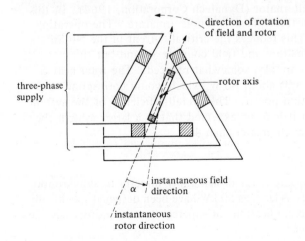

Figure 10.3.1. The superconductive synchronous induction motor.

operating current must be much less than the short sample current. This has the effect of increasing the a.c. armature losses.

10.3.4 Applications

The superconducting alternator is a device of low volume and weight, when considered in isolation from its refrigerator. Its efficiency is not very high and it is rather inflexible in operation, being sensitive to faults and overloads. The low volume and weight of the superconducting alternator make it attractive for aero-space applications, although considerable reduction in refrigeration requirements from present standards will be needed before an integrated alternator refrigerator system is competitive with conventional units.

In view of the overall inefficiency of the superconducting alternator–refrigerator system, it is unlikely that it will find many non-space applications. In the form of an integrated transmission system the superconducting alternator may find application, although it is more likely that the hybrid alternator with superconducting rotor and conventional stator will be first to find terrestrial uses.

10.4 Superconducting transformers†

The application of novel techniques to tasks of engineering requires a clear understanding of the disadvantages accruing from existing techniques that can be improved upon.

The conventional transformer has few disadvantages that can be significantly improved by the use of superconductors. In the high-power range, above 1 MVA, the inefficiency is very low, seldom amounting to more than 1% of rated power. The use of type I superconductors can, in theory at least, reduce these losses still further although not to zero; it is, however, doubtful whether the net running cost of a superconducting power transformer would be less than that of a conventional unit.

The other aspect of the power transformer which can be improved upon is its size. In view of its essentially passive function of transferring power between networks it is extremely bulky. The use of superconductors may solve the problems associated with size and weight. One other purpose in developing superconducting transformers is to produce a cryogenically efficient link between a conventional power network and a superconducting power transmission line.

The property of superconductors which is unique and which must be exploited is their ability to sustain high current densities. This may allow the construction of transformers having small core and small winding volumes. The superconducting power transformer may, under suitable conditions, use either type I or hard, type II superconductors.

† Wilkinson (1963).

10.4.1 Type I windings

Type I superconductors (or for that matter type II below H_{c1}) can carry lossless a.c. surface currents up to a limiting value given by Equation (10.2.1), provided that the ambient external field strength be no more than a few hundred amperes per centimetre. Now, although the peak flux density in a transformer core is around $1 \cdot 5$ T, the magnetising field strength at the windings is only about 10–100 A cm^{-1}. Thus, by retaining the iron core of a transformer, type I superconductors can be used in the windings.

The retention of the iron core poses a problem in cryogenics. The hysteresis and eddy power losses in the core dissipate about $\frac{1}{2}\%$ of the rated power in a transformer. These losses are roughly constant for varying temperature and do not decrease significantly at liquid helium temperatures. For instance the core loss of a 1 MVA transformer is about 500 W. At $4 \cdot 2°$K this would require $\frac{1}{2}$ MW of refrigeration. The core of a superconductive transformer must clearly remain at room temperature. This necessitates a non-conductive Dewar for the windings: at least the Dewar must not provide a continuous electrical path linking the transformer core. Plastic Dewars are now being made.

The very high current densities of type I superconductors are confined to the surface sheath, about 10^{-5} cm in depth. Consequently, if advantage is to be taken of these high current densities, conductors of very thin section must be used.

When power flow occurs between windings of a transformer field strengths considerably in excess of 100 A cm^{-1} are generated in each winding. The net magnetising field experienced by the core remains constant because the effects of primary and secondary windings cancel. Locally within the windings, however, fields greater than 1 kA cm^{-1} can be generated unless the windings are interleaved. This latter arrangement is obviously mandatory if type I superconductors are to be used. Unfortunately a consequence of the interleaved windings is that a great deal of space is occupied by major insulation which must now separate each layer rather than the complete windings. There is also increased dielectric loss as a result. Furthermore, the effect of local fields due to power currents is augmented considerably during overloads.

Present conclusions about the superconductive transformer with type I windings are that it is not an economic substitute for a conventional power transformer, even if the latter has to be built piecemeal on site. Dielectric losses, lead losses, and other heat inputs at low temperature would require considerable refrigeration power, and the windings would not be much more compact than copper windings because of the need for major insulation within the windings.

10.4.2 Type II windings

The use of windings of hard, type II superconductor in a transformer eliminates the need for interleaved windings, thereby relieving many problems of high-voltage insulation. However, power loss now occurs in the windings and the efficiency of the transformer decreases.

Simple calculations show that the iron core cannot be omitted from a transformer using type II superconducting windings. The omission of the iron core could, in theory, be compensated by increasing the number of winding turns, thereby limiting the magnetising current to the level that would obtain in a conventional transformer. But the amount of superconductor would be so great that the a.c. losses would become insupportable and the cost of material could not be economically justified.

Even with an iron core, a type II superconductive transformer could not achieve the overall efficiency of a conventional unit. A potential reduction in size of low-voltage units may be possible using superconductors. The volume of a transformer winding is accounted for by conductor, insulation, core, and support structure. In the high-voltage transformer, insulation accounts for a large part of the winding volume; consequently here the use of superconductive windings carrying high current densities would not achieve a worthwhile reduction in size. However, in low-voltage windings these high current densities could effect a decrease in size of the windings. It is quite improbable that, at present, the use of super-conductors in power transformers would prove to be economically advantageous.

10.4.3 Hybrid transformers

Electrically the transformer serves the purpose of changing impedance levels and giving electrical isolation between systems. The hybrid transformer can also provide thermal insulation between systems, which may be of value where conventional and superconducting transmission lines join.

The hybrid transformer as its name suggests consists of an iron core, a set of conventional windings, and a set of superconducting windings. Its main purpose would be to transfer power between room-temperature and cryogenic systems with a minimum heat input to the low-temperature region. It could find application to power systems in connection with a superconductive transmission line (see next section).

The most severe limitation of superconductive transformers, as of alternators, is their intolerance of overloads. Overload capacity can only be obtained in superconducting devices by installing more superconductor: this capacity is therefore an expensive characteristic of a superconductive transformer. Furthermore, the a.c. loss of a hard, type II superconductor increases as its critical current I_c. Since this loss scales directly as I_c^2 [see Equation (10.2.23)] overload capacity incurs a rapidly rising increase in power loss and therefore in inefficiency.

Despite the generally discouraging prospects for superconductive power transformers, small units up to a few kilowatts have been constructed.

One important application for superconductive transformers of up to 1 kW rating is in the flux pump (see Section 10.6).

10.5 Superconducting cables†

The application of superconductors to power transmission has received considerable attention in recent years as a result of the pressure to use cables in place of overhead transmission lines. The conventional cable suffers from two severe disadvantages—it is expensive and it has high capacitance. Its cost derives basically from the technical problem of the high-voltage insulation, coupled with cooling requirements, and the capacitance results from its geometry, coupled with its high operating voltage: a given capacitance is less embarrassing at lower-voltage levels. The capacitance in fact gives rise to high charging currents which flow in resistive elements of a power grid and which call for large reactive power outputs from generators.

Superconductors offer the possibility of a.c. or d.c. transmission at high currents and low voltages. The use of low voltages reduces the undesirable effects of capacitance in the a.c. system. Furthermore, of course, super-conductors offer the possibility of comparatively low losses in both the a.c. or d.c. transmission system. (The electrical losses may indeed be zero in some systems, but the cryogenic losses, heat input to the low-temperature region, must be included in the overall power balance.)

Two basic types of superconductive cable for power transmission are being investigated: the d.c. cable using hard, type II superconductors and the a.c. cable using type I superconductors (or type II below H_{c1}). We shall do no more here than state the principles and general characteristics of each type.

10.5.1 d.c. cable

If alternating currents or fields are not present, a hard, type II super-conductor will not experience hysteresis losses. Then large sections of superconductive material can be used and high currents can be carried. A practical cable for power transmission must be able to handle at least 10 MW of power, and more if distances are great. If we take this figure by way of example, a suitable 'cable' might consist of two fully stabilised composite conductors each of 1 cm² in cross-sectional area. The current would be 10000 A and the voltage 1000 V. The spacing between conductors would be about 1 cm. The force of repulsion between the conductors would be 1000 N m⁻¹ length. The operating voltage could easily be increased to 10 kV or more, thus making this type of cable capable of transmitting over 100 MW.

The heat input to the low-temperature region through radiation and support structure would be about 10 W km⁻¹. The heat input through the transition leads between room-temperature and the low-temperature

† Comm. I, Intern. Inst. Refrig. (1969).

environment at each end of the cable would be 20 W. If the d.c. cable were to form part of an a.c. system, conversion equipment would be required at each end.

10.5.2 a.c. cable

Unless almost overwhelming hysteresis losses are to be handled, only the use of superconductive skin currents can be contemplated in an a.c. cable. This implies the use of type I superconductors at low field strengths (or of course type II superconductors below H_{c1}).

Possible materials are lead or niobium deposited in a state of very high purity and free of strain on copper or other substrates. (This type of superconducting film is also used in the construction of high-Q cavities; see Section 17.2.)

Considering the application of niobium to an a.c. cable we note that its lower critical field H_{c1} is about $1 \cdot 6$ kA cm^{-1}. Therefore in an a.c. superconductive cable the peak incident field would be limited to about 800 A cm^{-1} of periphery. If we consider the ideal geometry of a coaxial system, the diameter of the inner conductor must be at least 3 cm if a current of 10 000 A is to be carried.

It can be seen therefore that the a.c. cable utilising superconductive skin currents will, in general, be bulkier than the d.c. counterpart, for a given power level.

The problems associated with the cooling of long lengths of a super-conducting cable would be difficult and the installation of the first commercial superconductive power transmission line is not likely to take place soon.

10.6 Flux pumps†

The considerable loss rate of liquid helium from a Dewar caused by the introduction of high-current leads has prompted the construction of a variety of flux pumps.

The flux pump is a device whereby power can be introduced either mechanically or electrically with small heat input to a low-temperature region and there converted to a low-voltage high-current supply.

Two types of flux pump have been developed to energise high-current superconducting magnets; they are the homopolar generator and the transformer–rectifier unit.

10.6.1 The homopolar generator‡

Figure 10.6.1 illustrates the principle of operation of the homopolar type of flux pump. The coil to be energised is connected through very-low-resistance or superconducting joints to a thin plate of pure annealed niobium or lead. A magnetic field of strength greater than the critical field of the plate is applied to the plate by means of a powerful

† Comm. I, Intern. Inst. Refrig. (1969).   ‡ van Suchtelen and Volger (1965).

permanent magnet, quenching superconductivity locally, as shown. The 'flux spot' is moved across the plate, in which there is always a superconducting path, and into the circuit of the coil. Because the coil circuit is composed of wire having a high critical field, further movement of the magnet along the indicated path leaves the flux trapped in the coil circuit. If the magnet is moved continuously on a circular path, flux will be swept across the plate and into the coil circuit to be trapped in each cycle when the magnet reaches point A of Figure 10.6.1.

In this device power is transmitted to the low-temperature region mechanically. Work is done each cycle in moving the magnet away from the coil circuit.

A moving flux spot can also be generated electrically as shown in Figure 10.6.2. Suitably distributed coils supplied with a three-phase current will generate a flux spot moving linearly across the plate. The diodes shown in the figure indicate that the coils are energised by a current of only one polarity so that the flux swept across the plate is of only one polarity.

The effective voltage output of the homopolar flux pump is given by

$$V = f\phi , \qquad (10.6.1)$$

where V is the mean voltage applied to the load coil, ϕ is the flux swept across the plate, and f is the number of transits per second. For a number of practical reasons f cannot exceed about 50 Hz. Now the time taken to energise a coil is

$$T = \frac{LI}{V} , \qquad (10.6.2)$$

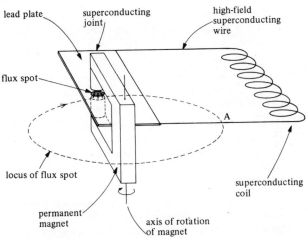

Figure 10.6.1. The basic homopolar flux pump. In this type of pump a flux spot is moved mechanically across a soft superconducting plate into a circuit of high-field hard superconductor.

where L is the coil inductance, I is the peak coil current, and V is the mean charging voltage. A 5 T, 15 cm bore coil might have a value of LI of about 1000 Vs. Thus to energise such a coil in 17 min, 1 V input is required.

Inserting this value of V in Equation (10.6.1)

$$\phi = 0 \cdot 02 \, \text{Wb.}$$

(10.6.3)

Since a maximum flux density of 1 T could be achieved in the flux spot, an area of 200 cm² would be required to generate 1 V. Although this could be done, the flux pump would be very large and rather inefficient.

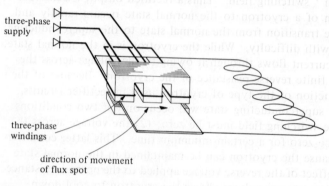

three-phase supply

three-phase windings

direction of movement of flux spot

Figure 10.6.2. The homopolar flux pump using a multi-phase supply to drive a flux spot across a superconducting plate.

10.6.2 The rectifier pump†

An alternative to the homopolar type of flux pump is the transformer–rectifier system, illustrated diagrammatically in all its stark simplicity in Figure 10.6.3.

The transformer has superconductive windings in both primary and secondary. The primary current is low so that the incoming leads from room temperature are small. The transformer core may be either of silicon iron or of a molybdenum permalloy: in both materials the hysteresis and eddy current losses are small and increase only slightly at low temperatures. In high-power flux pumps, the core can be thermally isolated from the low-temperature environment as discussed in Section 10.4.1.

The rectifying elements of this flux pump are a form of cryotron (see Section 14.2), in this instance of niobium which can be switched from the superconducting to the normal state by the application of a field of about 2400 A cm⁻¹. The resistance of a cryotron in the normal state is sufficient to limit the reverse current flow to a very small fraction of the

† Buchhold (1964).

forward current in the superconducting state. The mode of operation of the circuit as a whole is as follows.

An alternating voltage is applied to the primary winding. This induces alternating voltages of opposite phase in the two high-current windings. When the instantaneous voltage in one secondary circuit is of, say, negative polarity, the cryotron in that circuit is switched to the resistive state. The niobium foil in the other circuit is then in the superconducting state and a forward current flows in it. As the instantaneous secondary voltages reverse polarity, one cryotron reverts to the superconducting state and the other is switched to the resistive normal state by the application of the $2 \cdot 4$ A cm^{-1} switching field. Thus a rectified output is achieved.

The transition of a cryotron to the normal state occurs rapidly and reliably: but the transition from the normal state to the superconducting state is fraught with difficulty. While the cryotron is in the normal state, a small reverse current flows through it owing to the voltage across the load and to the finite reverse resistance of the cryotron. Because of the compact construction of the type of cryotron used in rectifier circuits, reversion to the superconducting state will occur only if two conditions are met: (1) the switching field must be zero; (2) the voltage across the cryotron must be zero for a certain minimum time. This latter condition is necessary because the cryotron can be maintained in the normal state by the heating effect of the reverse voltage applied to the normal resistance. A short dwell time at zero voltage allows the cryotron to cool down.

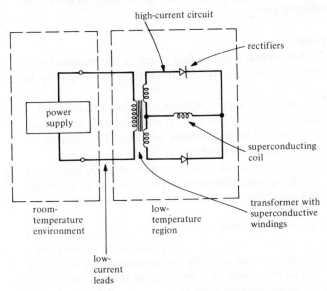

Figure 10.6.3. The basic circuit of the transformer–rectifier type of flux pump. The primary circuit carries a low current whilst the secondary circuit carries the high load current.

This dwell time can be achieved in two ways. Saturable reactors can be inserted in series in each secondary circuit to hold the voltage across the cryotrons nearly zero for a short time. Alternatively an input voltage waveform can be used which gives short periods of zero voltage.

The circuit configuration including saturable reactors has been developed quite intensively. It is shown in Figure 10.6.4.

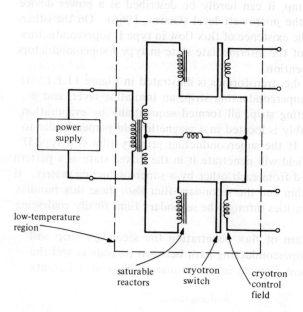

Figure 10.6.4. The complete circuit of the transformer–rectifier flux pump including saturable reactors.

Aggregation of flux in superconductors

11.1 The d.c. transformer and other effects of flux flow

The d.c. transformer has been allocated its own section in this chapter because, although its principle of operation is essentially similar to that of the homopolar flux pump, it can hardly be described as a power device because it operates at the microwatt level (Giaver, 1966). On the other hand, it demonstrates the existence of flux flow in type II superconductors and also the existence of the intermediate state in type I superconductors which deserves some mention.

The principle of the d.c. transformer is illustrated in Figure 11.1.1. It consists of a primary superconducting strip, an insulating layer, and a secondary superconducting strip, all formed sequentially by evaporation techniques. This assembly is located in a magnetic field perpendicular to the plane of the strip. If the superconducting primary film is a type II material, the ambient field will penetrate it in the mixed state as a pattern of flux quanta separated from each other by a superconducting matrix. If the insulating layer is thin and the secondary film also, these flux bundles will persist as discrete entities through the secondary film, finally coalescing some distance from it.

Because of this pattern of flux penetration the secondary strip will contain a continuous superconducting path between its ends as will the primary strip. A current flowing in the primary will exert a Lorentz

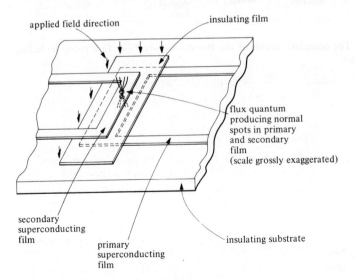

applied field direction   insulating film

flux quantum producing normal spots in primary and secondary film (scale grossly exaggerated)

secondary superconducting film

primary superconducting film

insulating substrate

Figure 11.1.1. The principle of the superconductive d.c. transformer. Flux bundles, isolated within vortices in the primary film, are swept by the Lorentz force across the secondary film.

force on the flux bundles, thereby driving them across both primary and secondary strips. A voltage will be generated in the secondary circuit in proportion to the rate at which the flux bundles cross it. Thus with the secondary winding open-circuited, the device is a current-to-voltage converter. In this configuration, the d.c. transformer is quite similar to the homopolar flux pump with a high impedance load (a magnet coil, for instance).

If the secondary circuit is wholly superconducting, the device will act as a d.c. transformer. The flow of flux bundles across the secondary circuit will proceed until the secondary current has risen to a level at which it balances the Lorentz force on the flux bundles due to the primary current.

The driving force on the flux bundles arises from the transport current flowing in the strip. It was mentioned in Section 8.1 that flux flow in a type II superconductor caused by a Lorentz force results in the eventual annihilation of the vortices (flux bundles) at the edge of the superconductor. This is what happens in the d.c. transformer if the secondary is not part of a superconducting loop. If, however, the flux bundles find themselves in a closed superconducting circuit after being swept across the primary strip they will be conserved.

Current transformation rates other than unity can be obtained by evaporating a number of separate secondary strips onto an insulated primary strip in such a way that the secondary strips are connected in series.

The superconducting films of a d.c. transformer may be of either type II or type I superconductor. The reason that flux flow takes place in a thin film of type I superconductor is that in a magnetic field the intermediate state is formed. The intermediate state allows the penetration of field into a superconductor whilst retaining a superconducting matrix for the flow of transport current.

11.2 The intermediate state of type I superconductors†

The intermediate state of a type I superconductor (not to be confused with the mixed state of a type II superconductor) arises during the penetration of flux into a body with a finite demagnetising coefficient. Consider, for instance, the penetration of flux into a thin strip of type I superconductor such as shown in Figure 11.2.1. Because of the shape of the strip and its basic diamagnetism, the external field is distorted so that the field strength at the edge of the strip is greater than the ambient field strength well away from the strip. If this enhanced field strength is less than the bulk critical field of the strip, the flux will not penetrate. This situation is illustrated in Figure 11.2.1a. If the ambient field strength is raised to the point at which the enhanced field strength at the edge of the strip is greater than the bulk critical field, penetration will occur. As a result of this penetration the field strength at the edge of the strip is

† Schoenberg (1952).

reduced to just less than the critical field and penetration then ceases until the ambient field is again raised. Thus, flux penetration into a type I superconductor of finite demagnetising coefficient proceeds in steps and a flux pattern of the kind shown in Figure 11.2.1b is set up in the superconductor. Such a pattern of normal and superconducting regions is called the intermediate state.

It is indeed possible to see the pattern of flux spots entering or leaving a superconductor in the intermediate state. Several experiments have been performed on discs of type I and type II superconductors including indium and niobium oriented perpendicular to a magnetic field (deSorbo and Healy, 1964; Schoenberg, 1952, p.95 ff.). The demagnetising effect of the disc causes the increased field at the edge of the disc to penetrate before the ambient field strength has reached the bulk critical field of the material. The flux spots can be made visible with the aid of a magnetically sensitive glass, such as a glass loaded with cerium phosphate which exhibits a strong Faraday effect. This effect is the rotation of the plane of polarisation of light passing through a magnetic field.

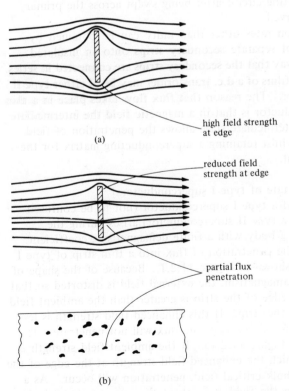

(a)

high field strength
at edge

reduced field
strength at edge

partial flux
penetration

(b)

Figure 11.2.1. The formation of the intermediate state in a type I superconductor having a finite demagnetising coefficient (see text).

Briefly, the polished disc of the superconductor to be studied is oriented in a Dewar perpendicular to the field of a solenoid. A thin piece of cerium phosphate glass is placed over the disc and illuminated by polarised light. The light reflected from the polished surface of the disc is viewed through a polarising analyser. In this way regions of the disc into which flux has penetrated show up in contrast with the superconducting regions. It should perhaps be noted here that the direct observation of flux movement is possible only when flux bundles containing at least 10^6 flux quanta are present. Even then the optical resolution obtained is usually not very high. The dissociation of flux into bundles in a superconductor in the intermediate state is in no way related to the mixed state of a type II superconductor. The latter is an intrinsic property of the superconductor. The intermediate state is associated with its geometrical characteristics.

It is of course also possible to observe statically the pattern of flux penetration in the intermediate state by the evaporation process described in Section 6.1.

Quantum devices

12.1 The quantum magnetometer†

Although the exploitation of the properties of zero resistance in super-conducting magnets is by no means unsophisticated, the use of the quantised fluxoid in a variety of quantum-measuring devices is perhaps one of the most esoteric applications of basic physical principles to be found at the moment.

In Section 6.1 we derived an expression for the energy of a super-conducting cylinder as

$$\epsilon = C\left(\pi r^2 H_e - \frac{N\phi_0}{\mu_0}\right)^2 . \tag{12.1.1}$$

We further saw that, as the field strength was varied, flux quanta moved into or out of the cylinder. Equation (6.1.33) shows that this movement of flux occurs at intervals of external field strength given by

$$\Delta H_e = \phi_0 (\mu_0 \pi r^2)^{-1} , \tag{12.1.2}$$

and Figure 6.1.4 shows the variation of the energy as a function of external field strength. The periodic sudden penetration of a flux quantum changes the energy and generates a voltage in the cylinder which may be inductively coupled to a circuit surrounding the cylinder.

The periodic variation in flux within the cylinder may be represented by a Fourier series of which the fundamental has the form

$$\Delta\phi_1 = \phi_0 \sin\left(\frac{2\pi H_e}{\Delta H_e}\right) , \tag{12.1.3}$$

where ΔH_e is the increment in external field strength between the movement of flux quanta.

Suppose now that the external field H_e consists of a slowly varying field H_{dc} and an oscillating component of peak amplitude H_{ac}.

Then the variation of flux within the cylinder may be written

$$\Delta\phi_1 = \phi_0 \sin\left\{\frac{2\pi}{\Delta H_e}[H_{ac}\sin(\omega t) + H_{dc}]\right\} , \tag{12.1.4}$$

where ω is the angular frequency of the oscillating field. The voltage induced by this fundamental flux variation is given by

$$e_1 = \frac{d}{dt}(\Delta\phi_1) = \frac{\phi_0 2\pi H_{ac}}{\Delta H_e}\omega\cos(\omega t)\left\{\cos\left[\frac{2\pi H_{ac}}{\Delta H_e}\sin(\omega t)\right]\cos\left(\frac{2\pi H_{dc}}{\Delta H_e}\right)\right.$$
$$\left. - \sin\left[\frac{2\pi H_{ac}}{\Delta H_e}\sin(\omega t)\right]\sin\left(\frac{2\pi H_{dc}}{\Delta H_e}\right)\right\} . \tag{12.1.5}$$

† Deaver and Goree (1967).

Before completing this analysis we must consider the form a quantum magnetometer might take.

A practical embodiment of a quantum magnetometer is shown in Figure 12.1.1 (Mercereau, 1967). A film of lead or tin is evaporated onto a glass rod of about 1 mm in diameter, forming a cylinder about 2 mm long and about 10^{-6} cm thick. By photoetching or other techniques the film is reduced to a very narrow neck, approximately 1 μm wide at one point. This cylinder is surrounded by a small resonant tank circuit, consisting of a coil and capacitor, which is energised at 10–30 MHz. The amplitude of the signal in the tank is measured after rectification and its variation plotted, or displayed directly, as a function of the d.c. and a.c. field strengths. Typical results are shown in Figure 12.1.2.

How these waveforms arise can be understood with reference to Equation (12.1.5) for the induced voltage around the cylinder. The tank circuit is coupled weakly to the superconducting cylinder, so that the rectified signal across the tank circuit consists of a constant term and a term proportional to the mean value of the voltage around the cylinder. From Equation (12.1.5) the latter is found to be

$$E = cJ_0\left(\frac{2\pi H_{ac}}{\Delta H_e}\right)\cos\left(\frac{2\pi H_{dc}}{\Delta H_e}\right) ,\tag{12.1.6}$$

where J_0 is the Bessel function of zero order and c is a constant.

The value of the constant is of no importance because it is only the period of the output signal, expressed in terms of H_{ac} and H_{dc}, with which we are concerned.

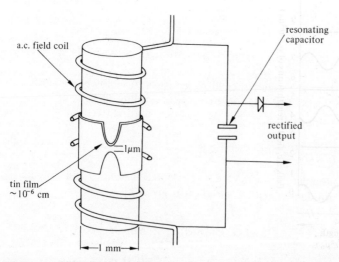

Figure 12.1.1. A thin-film quantum magnetometer. Flux quanta move into or out of the cylinder through the narrow neck. The change in flux generates a signal in the surrounding coil.

Equation (12.1.6) shows how the curves of Figure 12.1.2 must be interpreted. First, it will be seen that the presence of the Bessel function in the expression will lead to zeros in the variable part of the output signal, irrespective of the d.c. field. These null points will occur when

$$\frac{2\pi H_{ac}}{\Delta H_e} = 2\cdot405, \; 5\cdot520, \; 8\cdot654 \; ,$$

and other values of the argument for which J_0 is zero. For large values of the argument the spacing of the zeros is π. That is no output signal variable in d.c. field strength is obtained when

$$\frac{2\pi H_{dc}}{\Delta H_e} = N\pi \; , \tag{12.1.7}$$

Between these zero points the portion of the output signal, variable in H_{dc}, is a maximum.

Because the device is to be used as a quantum magnetometer it is usual to adjust the peak a.c. field strength so that

$$\frac{2\pi H_{ac}}{\Delta H_e} = 2\cdot405 \; ,$$

that is

$$H_{ac} = \frac{2\cdot405}{2\pi} \frac{\phi_0}{\mu_0 \pi r^2} \; . \tag{12.1.8}$$

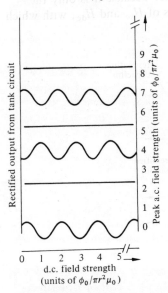

Figure 12.1.2. The rectified voltage across the tank circuit of the quantum magnetometer. This voltage is a function of the a.c. field (ordinates) and the d.c. field (abscissae) at the cylinder. There is no common base for the ordinates; the traces have been displaced successively upwards to give separation for clarity.

Putting $r \approx 1$ mm, $\mu_0 = 4\pi \times 10^{-7}$ H m^{-1} and $\phi_0 = 2 \cdot 07 \times 10^{-15}$ Wb then we find $H_{ac} \approx 0 \cdot 2 \times 10^{-3}$ A m^{-1}, which is of course obtainable with very simple circuits.

The cosine term in Equation (12.1.6) indicates that the variable part of the rectified output will have a periodic dependence on the d.c. magnetic field strength. It is this periodicity which is exploited in the quantum magnetometer to measure magnetic field strength and other parameters. The period is given by

$$\Delta H_{dc} = \tfrac{1}{2} \frac{\phi_0}{\mu_0 \pi r^2} , \qquad (12.1.9)$$

which for the values given above for r, ϕ_0, and μ_0, gives

$$\Delta H_{dc} = 0 \cdot 25 \times 10^{-3} \text{ A m}^{-1} .$$

The periodic variation of the rectified signal can be interpolated to at least 10% so that field strengths as low as 25 μA m^{-1} can be measured by this device.

The size of the narrow neck in the film is dictated by the speed of movement of a flux quantum through the film. The transit time for the flux quantum must clearly be less than the period of the a.c. field. The transit time will typically be given by the magnetic diffusivity multiplied by the width and the thickness of the neck. Thus

$$\tau \approx \frac{\mu_0 \omega d}{\rho} \text{ s.} \qquad (12.1.10)$$

For a lead film the normal resistivity will be about 10^{-8} Ω cm. Thus for a narrow neck of width 1 μm and thickness 10^{-6} cm the transit time will be 10^{-10} s, which is much shorter than the period of the a.c. field.

Although this type of quantum magnetometer is capable of measuring very small field strengths very accurately, it is less suitable for the measurement of high field strengths. A 10^{-6} cm lead film will be superconducting in field strengths up to a level given by Equation (7.1.6) of Section 7.1, that is

$$H_{ec} = 64 \cdot 0(3)^{\frac{1}{2}} \frac{0 \cdot 5 \times 10^{-5}}{10^{-6}} \approx 5 \cdot 5 \text{ kA cm}^{-1} .$$

However, the magnetometer will cease to operate long before such a field strength is reached because of electrical noise. This noise arises from the movement of the flux quanta in the film other than through the narrow neck.

The magnetometer can also be used as an absolute ammeter of low impedance. Suppose the cylinder is surrounded by a coil of fine superconducting wire of 100 turns cm^{-1}. A current flowing in this coil will generate a field given by

$$H = 10^4 I \text{ A m}^{-1} .$$

The magnetometer will measure this field with an accuracy of better than $0 \cdot 25 \times 10^{-3}$ A m^{-1}. In fact the rectified output from the magnetometer can probably be interpolated to $0 \cdot 025 \times 10^{-3}$ A m^{-1}. Then, as an ammeter, the device has an accuracy given by

$$10^4 I = 0 \cdot 025 \times 10^{-3} \text{ A},$$

$$I = 2 \cdot 5 \times 10^{-9} \text{ A}.$$

Although this accuracy is only comparable with standard measuring devices, the quantum ammeter has essentially no resistance. Another type of quantum ammeter using the Josephson junction is described in the following section.

Josephson junctions

13.1 The Josephson effects†

It will have been appreciated that only some of the aspects of the behaviour of superconductors can be explained anywhere near plausibly by means of a classical model of the electronic structure of a metal. Flux quantisation and superconductive tunnelling both require, for their full explanation, recourse to the grey abstraction of wave mechanics, although even in those cases classical analogies help to give conviction to the imagination. Similarly in Section 7.3 it was mentioned that Ginsburg and Landau analysed the spatial and temporal variation of the order parameter (density of superconducting electron pairs) to obtain a characterisation of a superconductor in terms of the parameter κ. This analytical *tour de force* invokes wave mechanics to describe the order parameter as a complex wave function; we were able to circumvent this analysis by considering the energy of a laminated superconductor without regard to the detailed variation of the order parameter. In deriving the conditions for the quantisation of flux in Section 6.1 we invoked only the simple principle that, if a particle moves without energy loss in a closed orbit, the probability of finding it at any point of the orbital path remains constant in time, that is its wavefunction is a standing wave.

To understand the Josephson effects we must closely re-examine the implications of the wave functions of superconducting electron pairs. We must start by accepting that a superconducting electron pair has a wave-function which describes the probability of finding the pair at a particular region of the superconductor at a particular time. Because the electrons of a pair may be separated by as much as the coherence length, we consider their centre of mass to be the point which is defined by the wavefunction. The wavefunction can be written as a complex function in which the spatial coordinates are real and the time coordinate is imaginary. The spatial wavelength of the wavefunction is related to the canonical momentum by the expression

$$\lambda = \frac{h}{p},$$

where λ is the wavelength, p is the canonical momentum, and h is Planck's constant. This is the fundamental relationship of de Broglie; it was introduced in Section 6.1. In a superconductor of no matter what size the canonical momentum of all electron pairs is the same, this being the condition derived in the BCS theory for the appearance of the large energy gap through collective interaction of the pairs. In terms of the wavefunction this commonality of momentum is equivalent to an identical

† Langenberg *et al.* (1966).

wavelength for the wavefunctions of all pairs throughout the superconductor. This is referred to as the 'long-range order' of the superconducting wavefunction.

It is worth pointing out here why we always refer to canonical
momentum. Consider a type I superconductor of large dimensions in a
magnetic field. The electron pairs at the surface are given a velocity and
hence mechanical momentum by the flux penetrating slightly into the superconductor. Deep inside the bulk of the superconductor the pairs have no
velocity. Thus, it might seem that the requirement of a common momentum
has been immediately violated. However, the pairs at the surface have
magnetic, as well as mechanical, momentum and these two terms cancel
everywhere so that the canonical momentum [see Equation (6.1.8)] of all
electron pairs in the superconductor is zero in this case. If the superconductor had a hole in it containing some flux quanta, the canonical
momentum of each pair would then correspond to the number of quanta.
The spatial wavelength of the wavefunction of the electron pairs circulating
around an integral number of flux quanta is, as we saw in Section 6.1, an
exact submultiple of the path length.

Now, as well as having a wavelength the wavefunctions of the pairs have
a phase in time: the phase of all electron pairs in a continuous superconductor is common. What this phase means in terms of the classical
picture of electrons is all but impossible to picture. However, in order to
have a model on which to focus the imagination, let us consider the
electron pairs very close to the surface of a superconductor. Because of
the presence of a very small field the electron pairs have a velocity, which
contributes in part to their momentum. We have already accepted that
the canonical momentum of all the pairs in the superconductor is the
same: consequently we can accept that the velocity of all the pairs at a
given distance from the surface is the same. [This arises because the field
strength and hence the magnetic contribution $(2eA)$ to the momentum is
the same.] Now let us picture the probability of finding one of these
electron pairs at a point and a time as a sinusoidal wave. (Actually the
probability is proportional to the square of the amplitude.) Figure 13.1.1
then shows the wavefunction of one electron pair as it moves just below
the surface of the superconductor. Now, at a point on the surface we
define a datum, and relative to this datum we can measure the temporal
distance of a point on the sinusoid. The number of cycles of spatial
variation of the wavefunction in this time is the phase of the superconducting electron pair. The rate of change of phase with time is a
measure of the energy of the pair. In a continuous superconductor the
wavelength of the wavefunctions of all electron pairs is the same and all
the wavefunctions advance at the same rate, thus maintaining the same
phase. If we were to draw the wavefunctions of other electron pairs on
Figure 13.1.1, the maxima of amplitude would not have to lie in the
same vertical positions but the wavelengths would all have to be the same

and they would all have to advance at the same rate. The datum is entirely arbitrary and therefore so is the phase. The absolute phase *per se* of an isolated superconductor has no meaning. All that can be said is that the phase of all electron pairs is the same. It must be remembered that we have picked out a layer in the superconductor in which all electron pairs have the same velocity. In a different layer the pairs will have a different velocity but *not* different canonical momentum; phase relates to canonical momentum, not simply to velocity, and our definition of phase in terms only of velocity is a simple expedient to aid the imagination. Let us examine further the implications of the phase of the wavefunctions.

The square of the amplitude of the wavefunction is proportional to the density of superconducting electron pairs. A difference in phase between the wavefunction of two sets of electron pairs therefore implies that, at an instant in time, there is a difference in electron densities of the two wavefunctions and therefore a flow of current.

Let us now consider two superconductors placed very close together but not in electrical contact. First, we must recognise that, although all electron pairs have the same phase in any one superconductor, the phases in the two superconductors need not be the same. If a phase difference exists between the superconductors, a current of electron pairs will flow by means of tunnelling, such as was described in Section 3.2, from one superconductor to the other.

If we take the current as being proportional to the difference in magnitudes of the wavefunctions, we may write for the fundamental term of a complex series

$$j = j_0 \sin \phi \,, \tag{13.1.1}$$

where ϕ is the phase and j_0 is the maximum current density that may flow across the gap. j_0 is a function of the characteristics of the super-conductors and the width of the gap.

The phase difference may be controlled by magnetic flux within the

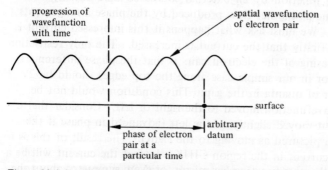

progression of wavefunction with time

spatial wavefunction of electron pair

surface

phase of electron pair at a particular time

arbitrary datum

Figure 13.1.1. The wavefunction of an electron pair close to the surface of a super-conductor. The spatial wavelength of the function is associated with the momentum of the pair: the time phase of the wave is associated with the energy of the pair.

gap or by a voltage difference between the superconductors. The control
of the phase by means of a magnetic field is called the d.c. Josephson
effect, and the influence of voltage difference on phase is the a.c.
Josephson effect. We shall consider first the d.c. effects.

13.2 The d.c. Josephson effect†

Let us suppose that a magnetic field H exists in the gap of width d and
length l between two superconductors. The total flux in the gap will be
$\mu_0 Hld$. The number of flux quanta in the gap will be

$$m = \frac{\mu_0 Hld}{\phi_0} \tag{13.2.1}$$

and m may be fractional. This may be contrasted with the requirement of
an integral number of flux quanta within a closed superconducting loop.
We saw in Section 6.1 that the phase change in the electron wavefunction
produced by a flux quantum is 2π. Then the phase change wrought by m
quanta is $2\pi m$.

Consider Figure 13.2.1a. This shows the gap between two superconductors containing a non-integral number of flux quanta. Around the
path ABCD the phase of the electron pair changes by 2π. The same
change occurs around paths DCEF and FEGH. Around GHJK the phase
change is less than 2π.

Now, in order to simplify the reasoning that follows, we shall assume
that the electron pairs in the lower superconductor have zero phase
everywhere and that all the phase change due to the magnetic flux in the
gap occurs in the upper superconductor. Then the phases of the electron
pairs can be as shown in Figure 13.2.1b. From Equation (13.1.1) it can
be inferred that a tunnel current will flow between the two superconductors, alternating in sign as the phase difference changes. These
currents are as shown in Figure 13.2.1c. It will be noticed that these
currents cancel in all regions except GHJK. Now suppose a current is
impressed on the junction by an external source so that initially the
current is the same as the net flow produced by the phase arrangements
of Figure 13.2.1. We must ask what happens if this impressed current is
varied. Suppose firstly that the current is increased. The only restriction
placed on the phasing of the electron pairs is that the phase difference
around the gap (or in our simple case along the top edge) should be 2π
times the number of quanta in the gap. This condition would not be
violated if the wavefunction moved to the right or left. Consider that of
Figure 13.2.1c but moved slightly to the left (advanced in phase if the
electron pairs are pictured as moving to the right). The result of this is to
increase the net current in the region GHJK. In fact the current will be a
maximum when the wavefunction has mirror, or even, symmetry about the

† Josephson (1964).

centre of the gap. The current will be zero when the symmetry is odd (see Figure 13.2.2b). Thus the impressed current will adjust the phase of the electron pairs so that the net tunnel current across the gap is equal to the impressed current, if this is possible. The limit is reached when even symmetry of the wavefunction is obtained.

The net maximum superconducting tunnel current can be obtained by integrating the area under the curve of even symmetry. Thus,

$$I = j_0 w \int_0^l \cos\left(2\pi m \frac{x}{l}\right) dx \ , \tag{13.2.2}$$

where l is the length of the junction and w is its width. ($2wl$ is then the surface area of the superconductors facing the gap.)

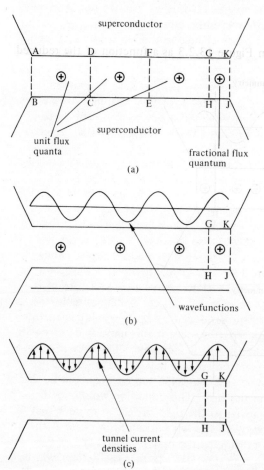

Figure 13.2.1. The wavefunctions and current density variations in a gap containing a non-integral number of quanta.

A maximum for I occurs when $m = 0$. This maximum is then given by

$$I_0 = j_0 wl .$$

Inserting this into Equation (13.2.2), the variation of the maximum superconducting tunnel current is found to be

$$I = I_0 \frac{\sin(2\pi m)}{2\pi m} . \tag{13.2.3}$$

$2\pi m$ is the total phase change due to the magnetic field. This can be rewritten in terms of the field strength H in the gap. Putting $H_0 = \phi_0/\mu_0 l d$ and using Equation (13.2.1), we get

$$I = I_0 \frac{H_0}{H} \sin \frac{H}{H_0} .$$

The modulus of I is plotted in Figure 13.2.3 as a function of the reduced

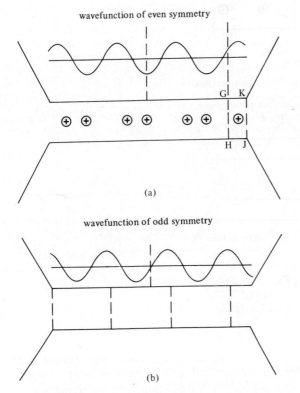

wavefunction of even symmetry

G K

H J

(a)

wavefunction of odd symmetry

(b)

Figure 13.2.2. The displacement of the wavefunctions by an impressed current. The maximum superconducting tunnel current is obtained by mirror symmetry as in the upper curve. Zero net current results from odd symmetry, as in the lower curve.

field H/H_0. The sign of I can be changed arbitrarily. The reversal of the direction of the impressed current will alter the position of the wavefunctions, but not the total phase difference around the flux, so that the net tunnel current matches the impressed current (Rowell, 1963).

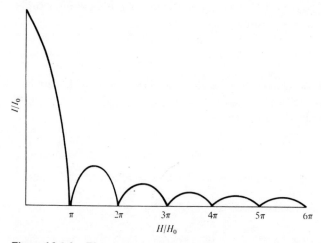

Figure 13.2.3. The variation of the maximum tunnel current in a Josephson junction as a function of the reduced field strength in the gap.

13.3. The a.c. Josephson effect†

The ability of the flux in the gap of a Josephson junction to change the phase, or energy, of the electron pair on either side of the junction must now be considered further. Magnetic flux is dimensionally equivalent to the product of potential difference and time.

Thus,

$$\phi = vt ,$$

where v is voltage and t is time. We should therefore expect that a discrete phase change could be produced by a voltage impulse, for this would have the dimensions of flux. Furthermore, a steady voltage should produce the same effect as a continually changing flux; and this effect would be a continually increasing phase. This in fact does occur in a Josephson junction.

Again consider Equation (13.1.1)

$$j = j_0 \sin\phi ,$$

ϕ being the change in phase. If ϕ is increasing uniformly with time, j will be an alternating function whose frequency therefore depends on the voltage across the junction. We may obtain the relationship between frequency and voltage from dimensional considerations alone. The frequency is in

† Josephson (1964).

fact equal to the equivalent rate of change of flux in quanta per second
(one flux quantum changes ϕ by 2π). Hence the frequency of oscillation
of the current density is given by

$$f = \frac{V}{\phi_0} .$$ (13.3.1)

This incredibly simple result is perhaps the most remarkable effect of
superconductivity. It relates voltage to frequency in terms of purely
physical constants. Because $\phi_0 = h/2e$, then

$$f = \frac{2eV}{h} ,$$ (13.3.2)

h being Planck's constant and e the electronic charge.

As an example, if a potential difference of 10^{-6} V is applied to a
Josephson junction, radiation is emitted (of minute intensity). The
frequency of this radiation is given by

$$f = \frac{10^{-6}}{2 \cdot 07 \times 10^{-14}} \approx 500 \text{ MHz} .$$

Conversely, if a Josephson junction is irradiated at 500 MHz, a potential
difference of 1 μV will appear across it in the presence of a net current
less than the maximum superconduction current at whatever the magnetic
field in the gap.

Although it is fairly simple to analyse the a.c. Josephson effect in purely
dimensional terms, it is extremely difficult to picture in detail what is
taking place in the junction, even in terms as concrete as the electron
pairs.

It must also be borne in mind that the voltage across a junction is not
in fact physically associated with a steadily increasing flux in the gap.
The only relationship between voltage and rate of change of flux is their
dimensional equivalence.

An alternative model for the effect of voltage on the phase of the
superconducting electron pairs is the following. As an electron pair tunnels
through the gap between the superconductors, its energy will tend to rise
(or fall) by 2 eV. On reaching the far side of the gap with this changed
energy the electron pair will find that it has the incorrect velocity to
match the correlated motion of the pairs. It will be therefore slowed
down (or accelerated) and in so doing will emit (or absorb) an energy
2 eV in the form of a radiated photon. The energy of a photon is related
to its frequency by the expression

$$\epsilon = hf ,$$

where h is Planck's constant and f is the frequency; hence,

$$2 \text{ eV} = hf .$$ (13.3.3)

as before.

13.4 Application of the Josephson effects

The main applications of the d.c. and a.c. Josephson effects lie in the area of the physics of superconductivity itself. However a laboratory galvanometer has been constructed, which exploits the ability of a weak superconducting link to carry a critical superconducting current and at the same time to admit flux (Clark, 1966). The device has many characteristics of a Josephson junction although the critical currents do not appear to decrease in amplitude as the enclosed flux increases in a way that would be expected of a Josephson junction.

The galvanometer is formed from a niobium wire around which a blob of solder is set. The oxide film on the niobium wire forms the gap between the two superconducting elements of the circuit. In fact the device behaves as if two weak links were formed between the solder and the niobium at each end of the solder. The arrangement and dimensions of the junction are shown in Figure 13.4.1.

In a Josephson junction with uniform gap the phase of the wavefunction is assumed to change uniformly along the face of the gap according to the enclosed flux in the gap. Furthermore, the variation of tunnel current density is assumed to be roughly sinusoidal; these assumptons are confirmed experimentally. However, in the present case the tunnel currents appear to flow only at the ends of the blob of solder. These two tunnel currents depend on the phase difference of the wavefunctions on each side of the gaps. As in the basic d.c. Josephson effect the critical superconduction

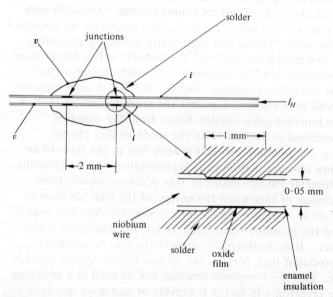

Figure 13.4.1. Arrangement and dimensions of a Josephson junction in use as a galvanometer.

current through the gaps is a periodic function of the flux in the gaps. This flux is directly proportional to the current in the niobium wire. Therefore the number of cycles through which the critical superconducting tunnel current has passed is a measure of the current in the wire. In the present device one flux quantum is equivalent to an increment of 200 μA in the wire. The device operates as follows.

A sensing current, oscillating at 20 kHz is introduced through the leads i–i. The amplitude of this current is such as to exceed the critical tunnel current at some point in the cycle of changing flux. During each cycle of sensing current a voltage is developed as the current increases through the critical value; the voltage disappears as the sensing current decreases through this value later in the cycle. Thus the mark–space ratio of the voltage waveform is a measure of the point in the repeating cycle of sensing current at which the critical current is passed.

Now, the main current in the niobium wire controls the flux in the gap and hence the critical tunnel current of the gap. If the main current changes slightly, the mark–space ratio of the voltage waveform will change. If the main current changes by such an amount that the flux in the gap changes by one-half of a flux quantum, the mark–space ratio of the sensing voltage will complete one cycle of variation. The measurement of the current in the wire therefore consists of counting the number of cycles of variation of the mark–space ratio of the sensing voltage and, for greater accuracy, of interpolating the mark–space ratio of the final cycle.

The device shown in Figure 13.4.1 has a sensitivity of about 1 μA using 1% interpolation of the cycle of critical tunnel current. Although such current sensitivity is not very great, the device has essentially no resistance and very low inductance. Hence it is particularly useful for measuring small voltages in low-impedance sources. An example is the measurement of the voltage generated in a low resistance by a small current. The inductance of the galvanometer circuit may be 10^{-8} H. A time constant in the source circuit of 1 s will thus permit the measurement of 10^{-14} V.

The Josephson junction galvanometer differs from the quantum magnetometer described in Section 12 in the mode of flux change. Because the Josephson junction contains a weak link in the form of an insulating gap, flux movement through it is continuous. In the thin-film quantum magnetometer the movement of flux is discontinuous, there being a sudden change of flux when the energy of the flim has risen to a critical value. As an ammeter the Josephson junction described here is less sensitive than the quantum magnetometer but its inductance is also considerably lower. It is, furthermore, somewhat easier to construct.

It may be appreciated that, because the critical tunnel current depends on the flux in the gap, the Josephson junction can be used as a switching element, like the cryotron. In fact it is capable of switching speeds of as little as 1 ns, and the device is being actively developed as a computer element.

14

The application of superconductivity to computers

14.1 Computer elements

Superconductors can exist in either of two distinct phase states, the normal (resistive) state or the superconducting (resistanceless) states. Since a duality of states typifies all logic elements used for computation in the binary code, it is consequential that superconductors are being actively considered as components of computing systems.

Type I superconductors possess the basic characteristics needed in a computing element, namely a duality of states, very fast switching speeds between states, small size, low power requirements, and low power loss. Two fundamental types of logic element exploiting these properties of superconductors are being developed: they are the cryotron and the Crowe cell. The cryotron is a switch and can be used to perform the algorithmic functions of a computer. The Crowe cell, or in its present form the continuous film memory, is a bit storage unit capable of a very high packing density.

14.2 The cryotron†

The cryotron was invented in 1956. It consists of two superconducting circuits, one of which is controlled by the other. Figure 14.2.1 shows the basic cryotron unit. A tantalum wire of $0 \cdot 009$ in diameter and a few inches long is wound with a single layer of insulated niobium wire of $0 \cdot 003$ in in diameter along a 1 in length. As the critical temperature of tantalum is $4 \cdot 48°K$, a very small magnetic field will drive it normal at $4 \cdot 2°K$. Such a field can be supplied by the niobium winding which itself has a much higher critical field at $4 \cdot 2°K$.

The tantalum wire is called the gate and the niobium winding the control. By Silsbee's rule the maximum permissible gate current is given by

$$I_g = \pi H_c d , \qquad\qquad (14.2.1)$$

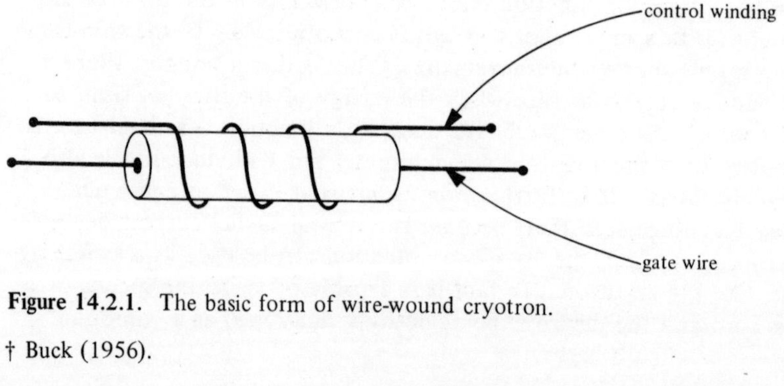

Figure 14.2.1. The basic form of wire-wound cryotron.

† Buck (1956).

where H_c is the critical field of the tantalum at $4\cdot2°K$ and d is the wire diameter.

The control current required to give this critical field is given by

$$I_c = \frac{H_c}{n} ,\tag{14.2.2}$$

where n is the number of turns of niobium wire per unit length. The current gain α of the cryotron is then

$$\alpha = \frac{I_g}{I_c} = \pi n d .\tag{14.2.3}$$

A cryotron with a current gain of more than unity can directly control another identical unit. Two such cryotrons with their control and gate circuits cross connected, as shown in Figure 14.2.2, form a bistable computing element. This is the basis of proposed counter and shift-register circuits. The cryotron can also be used in logic circuits.

The switching speed of a cryotron flip-flop (bistable pair) is determined largely by the inductance of the control winding and the resistance of the gate in the normal state. For fast switching the time constant given by

$$\tau = \frac{L_{control}}{R_{gate}}\tag{14.2.4}$$

must be small. The wire-type cryotron of Figure 14.2.1 has time constant τ of about 10^{-4} s. This is not fast enough for application in present computers. A modification which increases R is the substitution of a tantalum tube for the gate wire, as shown in Figure 14.2.3. Since the

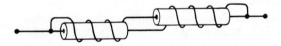

Figure 14.2.2. A bistable pair of cryotrons. Either cryotron can carry current, but current cannot be conducted simultaneously by both.

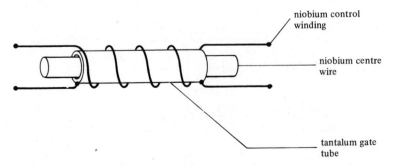

Figure 14.2.3. Modified wire-wound cryotron having low inductance.

gate current flows only in the surface of the tantalum, to a depth of about 10^{-5} cm, the interior bulk of the gate wire can be removed without affecting the critical gate current. The normal resistance of the gate will now be enormously increased.

To form a tube with a wall thickness of $\sim 10^{-5}$ cm requires a vacuum evaporation technique (which is incidentally best effected with metals other than niobium and tantalum). Suppose the gate tube is formed by evaporating a thin layer of an insulator onto a niobium wire and then by evaporating tantalum onto the insulation. This achieves two objectives, namely the formation of a thin gate element and the reduction of the inductance of the control winding. The cross-sectional area available for the accommodation of control flux is now the annular space bounded on the inside by the niobium wire into which the flux cannot penetrate. This area is evidently much smaller than the cross section of the original tantalum wire so that the control inductance is now very small.

The final stage of evolution of the cryotron comes with the realisation that the cylindrical geometry is inferior, from a production point of view, to a plane geometry and that a planar cryotron with a net current gain is easily achieved. It can be made by the sequential evaporation of tin, lead, and silicon monoxide layers through masks. With suitable photograph and masking techniques planar cryotrons can be deposited to a surface density of more than 100 units cm^{-2}.

Figure 14.2.4 shows the stages in the preparation of a typical evaporated-film cryotron flip-flop. It will be noticed that the control layer crosses the gate layer but obviously cannot form a coil. The criterion for a current gain in the planar cryotron is now that the width of the gate strip should be greater than that of the control strip. In practice a ratio of about 10 in widths is maintained.

The lead base plane serves the purpose of the niobium wire mentioned above, in that it reduces the area available to accommodate flux, thereby reducing the inductance of the control circuit. Silicon monoxide is used

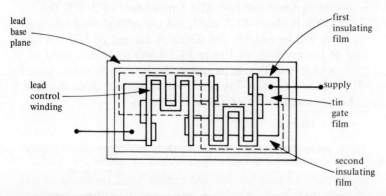

lead base plane

first insulating film

lead control winding

supply

tin gate film

second insulating film

Figure 14.2.4. Evaporated-film cryotron.

as an insulator between the layers of superconductors. In a typical evaporated-film cryotron the evaporated layers are about 1 μm in thickness; the width of the tin gate is about 400 μm and of the control 40 μm.

The inductance of the control strip is

$$L_c = \frac{h_{cB} W_g \mu_0}{W_c} , \qquad (14.2.5)$$

where W_g is the width of the gate strip, W_c is the width of the control, and h_{cB} is the height of the control strip above the base plane. The normal resistance of the gate is

$$R_g = \frac{\rho W_c}{W_g t_g} , \qquad (14.2.6)$$

where ρ is the normal resistivity of the tin in ohm cm and where t_g is the thickess of the gate strip. The switching time is then given by

$$\tau = \mu_0 h_{cB} t_g \alpha^2 \rho^{-1} , \qquad (14.2.7)$$

where $\alpha = W_g/W_c$, the current gain. Typically for the evaporated cryotron shown in Figure 14.2.4

t_g = 10^{-4} cm,
h_{cB} = 3×10^{-4} cm,
ρ = 2×10^{-7} ohm cm,
α = 10,

whence $\tau = 0 \cdot 188$ μs. Present trends towards decreased layer thickness indicate that this time constant is likely soon to be decreased by an order of magnitude. However, it serves to show that the evaporated-film cryotron is potentially a very fast computing element with a very high packing density.

The operating current of the cryotron is set by the critical field of the tin gate at the operating temperature. At a temperature of $3 \cdot 4°$K the bulk critical field of tin is about 40 A cm^{-1} and the critical field strength of a film 1 μm thick will be roughly the same. A current of 300 mA in the control strip of the cryotron of Figure 14.2.4 will generate a field of about 70 A cm^{-1}, which is sufficient to switch the gate. The same current in the gate will produce a field of 7 A cm^{-1}, which will not drive the gate normal. The energy stored in the cryotron is given by

$$E = \tfrac{1}{2} L_c I^2 \approx 20 \times 10^{-12} \text{ J} .$$

If all this energy is dissipated in the low-temperature environment, as seems probable, a bit rate of 10^8 s^{-1} will cause a dissipation of only 2 mW, which represents a negligible heat load. Input and output leads are likely to give rise to a much larger heat load. Cryotron stores and algorithmic systems containing up to 10^7 bits are being presently constructed and it seems probable that they will soon be incorporated into computers.

14.3 The Crowe cell†
Unlike the cryotron, the Crowe cell and its descendant, the continuous film memory, has no current gain. One Crowe cell cannot drive another except through a separate amplifier. The cryotron is, of course, capable of driving one of its own kind.

The Crowe cell in its original conception is shown in principle in Figure 14.3.1. A lead plate has two D-shaped holes in it on either side of a cross bar. Binary information is stored in the form of persistent currents flowing in the cross bar. Currents flowing in the direction shown in Figure 14.3.1 might be defined to represent the bit 1 and currents of the opposite bit 0. Two drive wires X and Y placed close to the cross bar as shown in the figure enable information to be 'written' and enable the information to be 'read' in conjunction with the sense wire. The drive and sense wires may be normal metal or superconductors.

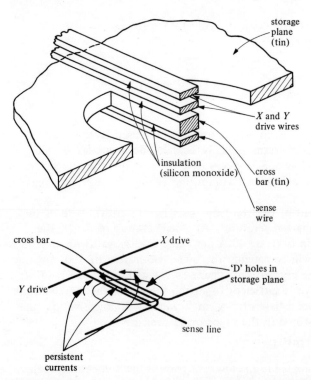

Figure 14.3.1. The basic configuration of the Crowe cell. The binary digits 1 or 0 can be identified with a persistent current flowing in one direction in the cross bar.

† Crowe (1957).

The 'writing' of information in the form of a '1' (positive persistent cross-bar current) or of a '0' is achieved when currents, together exceeding a certain minimum value, flow through the X and Y drive lines. Consider the waveforms of Figure 14.3.2a. The thin horizontal lines of the upper waveform refer to the critical current of the cross bar: the dashed waveform represents the cross-bar current and the solid line the combined net drive current normalised to minus the cross-bar current.

A rising current is applied to the X and Y drive: this induces a current in the cross bar of opposite sense. When the cross bar current has reached its critical value, the cross bar goes normal admitting flux to the hole. When the drive current is reduced, the cross-bar current falls, finally

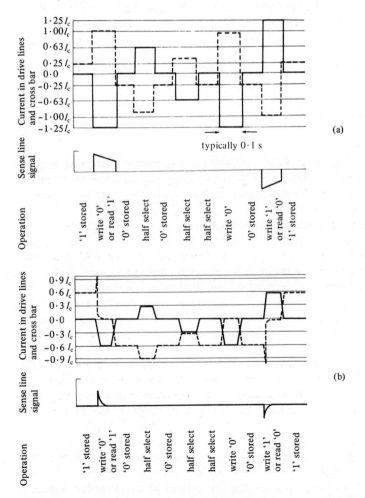

Figure 14.3.2. A series of 'read' and 'write' waveforms in a Crowe cell. Slow operation is represented by the waveforms in (a). Fast operation gives the waveforms of (b).

reversing its polarity and remaining as a persistent current after the drive current has become zero.

Now the X drive lines will traverse many units in a matrix of cells; so will the Y drive lines. Thus each cell must be unresponsive to half-select pulses, that is to currents in only one drive line.

This clearly implies that the stored persistent current plus the half-select current (drive current in one line only) must be less than the critical cross-bar current.

Thus

$$I_p + I_d < I_c ,\tag{14.3.1}$$

where I_p is the stored persistent current in the cross bar, I_d is the half-select (single-drive) current, and I_c is the cross-bar critical current. But,

$$I_p = 2I_d - I_c ,\tag{14.3.2}$$

so that from Equations (14.3.1) and (14.3.2)

$$I_d < \tfrac{2}{3} I_c .\tag{14.3.3}$$

To achieve a persistent current at all, however,

$$I_d > \tfrac{1}{2} I_c .\tag{14.3.4}$$

Equations (14.3.3) and (14.3.4) delineate a very narrow permitted range of drive currents; for this reason Crowe cells are difficult to operate in a large memory matrix where non-uniformity in the characteristics of the cells is likely to occur.

We have so far considered the storage of '1' and '0'. To read '1' a negative drive pulse is applied as shown in Figure 14.3.2 inducing a voltage pulse in the sense line. This process is destructive for now a '0' is stored. If we again read with a negative drive pulse, no voltage appears in the sense line, thus indicating '0' stored, but the read-out is now non-destructive. Thus, to return the cell to its normal state after reading '1' it is necessary to write '1'. After reading '0' the cell is still in its normal state. Similarly, write '1' is only effective if '0' is stored, and conversely, write '0' succeeds only if '1' is stored.

The reader may now rightly suspect that the waveforms of Figure 14.3.2a are of academic interest only. The time scale noted in this figure is absurdly slow for any practical device. Such a time scale allows the cross bar to quench and recover many times on attaining its critical current so that the persistent current is indeed given by Equation (14.3.2).

What happens when the time scale is reduced to microseconds is displayed in Figure 14.3.2b. Now, when the cross-bar current reaches the critical value, heating occurs in the cross bar which maintains it in the normal state. The time constant for the cooling of the cross bar is such that the cross bar does not recover its superconductivity until the flat top of the drive pulse has been reached. During this period the cross-bar current is zero so that when the drive currents are removed, the cross-bar current rises to minus the drive current.

Now, the limits on the drive currents are given by

$$4I_d > I_c$$
$$2I_d < I_c ,$$

so that

$$I_d \approx 0 \cdot 33 I_c .\qquad(14.3.5)$$

With this criterion the effects of read and write operations are as described above. The waveforms of a series of such operations are shown in Figure 14.3.2b.

A modern development of the Crowe cell is the continuous film memory (Burns *et al.*, 1961). To fabricate a simple matrix of memory cells it is an advantage to dispense with the holes on each side of the cross bar. This can be done by using a continuous film of tin or indium as the storage plate. As evaporated, these films contain a mass of holes and irregularities which serve to trap flux. The X and Y drive lines and the sense line now take the form of evaporated lead films and silicon monoxide is used as interlayer insulation. The problems of operating margin and uniformity are considerable but extensive work is being done on the continuous film memory (CFM) because of its high packing density, simplicity of fabrication, and speed. Operating speeds of up to 10 MHz should be possible.

The application of superconductivity to electronics

15.1 Circuit elements†

Superconductors are capable of serving amongst other things as amplifiers, detectors, oscillators, and multipliers.

A variety of properties of superconductors is employed in these devices. Both type I and type II superconductors can be pressed into service as electronic devices. As an example of the electronic applications of super-conductors we shall consider the detection of infrared radiation; this has been achieved in practice and represents a particularly appropriate application for superconductors.

15.2 The superconductive bolometer‡

The fundamental problem associated with the detection of very-low-intensity infrared radiation is that of noise. At room temperature an ideal emitting surface radiates at a rate of 0.05 W cm^{-2}, with a distribution of frequencies peaking at 10^{13} Hz. The limit to the sensitivity of a detector in the presence of this radiation flux is the noise associated with the random nature of the radiation. Clearly no change in radiation level smaller than these random variations can be interpreted usefully.

Detection in the visible and ultraviolet is not handicapped by room-temperature radiation. In this frequency range the photon energy is high and individual photons can be detected. The best detector in this range is probably the photomultiplier which has the greatest sensitivity of all radiation detectors but only to very-high-frequency photons, whose discrete energies are greater than the work function of the photocathode of the multiplier.

Clearly then, if high sensitivity is required in the infrared, detectors operating at low temperatures must be used. This temperature requirement suggests that superconductors might be successful radiation detectors if a strongly temperature-dependent property could be exploited. Such a property is the resistivity of a type I superconductor at its transition temperature. In Figure 15.2.1 the resistivity of pure tin in the region of its transition temperature is shown. If incident radiation can be made to change the temperature of a sample of tin by as little as 10^{-4} degK, the consequent change in resistance should easily be detectable. This principle has been applied successfully to a bolometer with a sensitivity of 10^{-12} W.

The general construction of the detection system of this bolometer is shown in Figure 15.2.2. The sensitive element is a tin film 2 mm $\times$ 3 mm $\times$ 3 μm thick evaporated onto a thin mica substrate. This element is supported from a comparatively heavy copper block by four nylon threads, each 10 μm in diameter and coated with 1 μm of lead,

---

† Chirlian (1962).   ‡ Martin and Bloor (1961).

deposited by evaporation. Other than through these threads the film is thermally isolated, being in a vacuum. In the superconducting state lead has a very low thermal conductivity but, of course, zero electrical resistance. Thus the tin film is in good electrical contact with an external bridge circuit but in relatively poor thermal contact with the block. Indeed the thermal time constant of the tin film is about 5 s. The block is equipped with a heater which forms part of a temperature-stabilisation circuit.

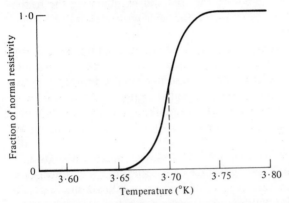

Figure 15.2.1. The resistivity of tin as a function of temperature close to the critical temperature.

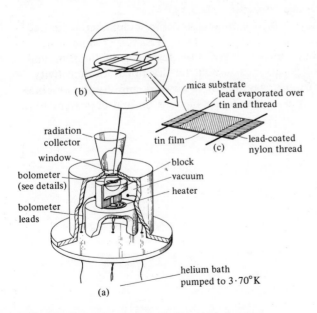

Figure 15.2.2. The superconducting tin bolometer.

In order to obtain high sensitivity from the superconducting bolometer, the temperature of the tin film, in the absence of radiation, must be maintained at the centre of the resistive transition region (see Figure 15.2.1). This necessitates control of temperature to 10^{-5} degK.

Such control is achieved in two steps. Firstly, the helium bath is pumped to a pressure of 480 torr; by sensing the vapour pressure and by regulating the speed at which the vapour is pumped off, the temperature of the bath is held to within 10^{-3} degK. The second stage of control is achieved by sensing the resistance of the film and by adjusting the heater current to maintain the temperature of the film with a stability of 10^{-5} degK.

It might seem that the temperature control circuits had eliminated any chance that radiation could affect the resistance of the film. However, the thermal time constant of the film as mounted on the heater block is some few seconds. Therefore, if the incident radiation is interrupted at a frequency of, say, 10 Hz, the temperature-stabilising circuit will not be able to respond to the changes in resistance of the tin film caused by the radiation.

The operation of the control circuit is as follows: incident radiation, modulated at 10 Hz, changes the resistance of the tin film which unbalances a bridge circuit. The imbalance signal, of frequency 800 Hz modulated at 10 Hz, is fed to an amplifier whose output is divided between two functions. It is rectified in synchronism with the 10 Hz modulation and it controls the temperature of the block. Only slow variation in the output of the amplifier affects the latter, however.

The ultimate sensitivity of the superconducting bolometer is limited by three types of noise present in the system. These are temperature fluctuations in the film, Johnson noise in the resistance of the film, and amplifier noise. For the particular bolometer described, the sensitivity was limited to 10^{-12} W, at a wavelength of about 1 mm, by the amplifier noise. The bolometer was probably capable of a tenfold improvement on this.

High frequency applications of superconductors

16.1 Frictionless bearings

Together with perpetual motion, the frictionless bearing has long been an object of humour in mechanics, to which might be added the d.c. transformer in electrical engineering. Quantum effects, however, now epitomise perpetual motion, in the motion of atomic electrons and macroscopically in the persistence of superconducting currents. Super-conductivity has produced the d.c. transformer and superconductivity can also be used to realise frictionless bearings. Indirectly, this is also due to quantum effects.

How the superconductive bearing works is not hard to see. Because a field less than the bulk critical field will not appreciably penetrate a type I superconductor, there is no resistance to the movement of magnetic flux past it. Even the small flux actually penetrating a type I superconductor represents a reversible effect which is therefore lossless (see Section 2.1).

The thrust that can be supported by a superconductive bearing is given in Newtons by

$$F = \tfrac{1}{2}\mu_0 q A H_c^2 , \qquad (16.1.1)$$

where H_c is the bulk critical field in amperes per metre, A is the area of bearing surface in square metres, and q is a factor, less than unity, which depends on the shape of the bearing and the gradient of the field. If the bearing uses niobium just below H_{c1}, the maximum thrust will be about 15000 N m^{-2}. The superconductive bearing is obviously appropriate for rotating superconductive machinery such as the alternator, but the only application it has so far seen is in the superconductive gyroscope.

16.2 The superconductive gyroscope†

In principle this device consists of a superconducting sphere supported freely in vacuum by the gradient of a magnetic field. The sphere is set spinning and the position of its poles of rotation give the required angular read-out. In principle it is an elegantly simple device but a variety of imperfections exist in a real system which cause precession, that is a systematic wandering of the poles of rotation.

Precession is caused by a torque applied to the spinning sphere at right angles to the axis of rotation, as shown in Figure 16.2.1. The rate of precession is given by

$$\omega_p = \frac{Tg}{I\omega} , \qquad (16.2.1)$$

where ω_p is the precessional speed in radians per second, ω is the spin

† Harding (1966).

rate in radians per second, T is the applied torque, I is the moment of inertia, and g is gravitational acceleration.

The main causes of this precessional torque are (1) non-sphericity of the superconducting sphere, (2) trapped field (although an ideal super-conductor in a field less than H_{c1} should show a complete Meissner effect, it is possible for a real material to trap magnetic flux if cooled in a magnetic field; such a field trapped in a gyroscope rotor would interact with normal metals in its vicinity to produce eddy current drag), and (3) inhomogeneity in the surface properties of the sphere. [If the local value of H_{c1} varies significantly over the surface of the sphere flux penetration may take place locally to cause a tangential retarding force which varies from point to point of the surface; such asymmetric forces will cause precession. A uniform penetration of flux (in excess of the London penetration) will cause a uniform drag with a consequent decrease in rotational speed accompanied by a rise in temperature but without precession.]

If precession were the only problem to be solved, a gyroscope rotor could be fabricated of any superconductor, reversible or hysteretic, provided only that it is uniform. However, drag is also an important consideration because not only does it directly reduce rotor speed, which affects precession through relation (16.2.1), but it also heats up the rotor and it is this that is likely first to limit the running time of the gyroscope. Consequently only very pure materials can be used. Furthermore, the maximum size of sphere that can be levitated depends on H_c or H_{c1}. For this reason and because of the advantage of an operational temperature of $4 \cdot 2^\circ K$, only lead and niobium can be considered. Of these, because of its low density and higher critical field, niobium is preferred; also, lead is mechanically less stable than niobium.

A further cause of temperature rise of the rotor is radiation. Were it not for the need to examine the rotor optically to determine the pole position, there would be no radiated heat input. This component of the heat input can, however, be reduced to a low level.

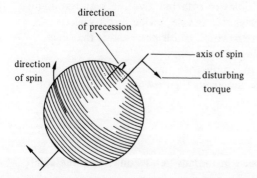

Figure 16.2.1. The cause and direction of precession in a gyroscope.

All these effects result from imperfections either in the fabrication of
the rotor or in the engineering concepts of the system, such as the read-
out system.   There exists also, however, a fundamental limitation to the
minimum precession of the rotor.   This is caused by the London moment
mentioned in Section 6.2.   To see how this arises consider Figure 16.2.2.
Firstly, the sphere is rotated with its spin axis in line with the axis of the
supporting field.   The London moment is manifest as a small field
symmetrical about this axis as shown and producing no net interaction
force with the suspension field.   Next, owing to movement of the support
coils caused by movement of the body in which the gyroscope is mounted,
the field configuration changes to that of Figure 16.2.2b, so that the spin
axis is out of line with the axis of the support field.   The London

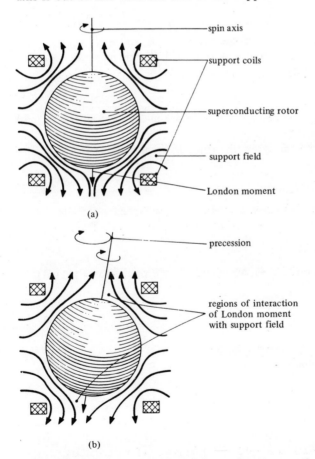

Figure 16.2.2 The London moment of a spinning superconductor. (a) The London
field is parallel to the external support field and no torque is produced. (b) The
London field is angled to the external field, resulting in a torque on the sphere. This
torque causes precession in the direction shown.

moment is still in line with the spin axis, however, so that it now interacts with the asymmetric component of the support field caused by the weight of the sphere to produce a torque tending to pull the spin axis into line with the axis of symmetry. This torque causes precession.

Since this torque represents a fundamental limit to the accuracy of the gyroscope, it is instructive to calculate its effect.

From Equation (6.2.6) the London field is given by

$$H_L = \frac{2m\omega}{\mu_0 e} , \tag{16.2.2}$$

and the magnetic moment per unit volume of the sphere is

$$M = \frac{2m\omega}{e} , \tag{16.2.3}$$

where ω is the angular velocity of the rotor, and m and e are the electronic mass and charge. In an external field H_e this magnetic moment causes a torque T given by

$$T = MH_e V , \tag{16.2.4}$$

where H_e is the net field at the rotor due to the asymmetry in the support fields caused by its weight.

Suppose the average field below the rotor is H_1 and above it is H_2. Then

$$\tfrac{1}{2}A\mu_0(H_1^2 - H_2^2) = \sigma Vg , \tag{16.2.5}$$

where A is the equatorial cross-sectional area of the sphere, V is its volume, σ is its density, and g is gravitational acceleration. Then

$$H_e = H_1 - H_2 = 2\sigma Vg[\mu_0 A(H_1 + H_2)]^{-1} . \tag{16.2.6}$$

If we combine (16.2.3), (16.2.4), and (16.2.6), the precessional velocity ω_p is given by

$$\omega_p = \frac{\mu_0 H_L H_e V}{I\omega}$$

$$= \frac{2(m/e)\omega[2\sigma V^2 g/\mu_0 A(H_1 + H_2)]}{\tfrac{8}{15}\pi\sigma a^5\omega} , \tag{16.2.7}$$

which simplifies to

$$\omega_p = \tfrac{40}{3}\frac{m}{e}g[\mu_0 a(H_1 + H_2)]^{-1} . \tag{16.2.8}$$

By way of example, consider a solid niobium rotor of 1 cm radius,

$$H_1 \approx H_2 = 800 \text{ A cm}^{-1} , \qquad \frac{e}{m} = 1\cdot 76 \times 10^{11} \text{ C kg}^{-1} ,$$

$$g = 9\cdot 81 \text{ m s}^{-2} ,$$

then,

$$\omega_p = 3 \cdot 72 \times 10^{-7} \text{ rad s}^{-1}$$
$$= 1 \cdot 84 \text{ deg day}^{-1}.$$

This precession rate is the limiting accuracy of a solid niobium rotor of 1 cm radius subject to gravitational acceleration. In zero-g conditions there is of course no precession due to the London moment. From Equation (16.2.8) it can be seen that, on Earth, ω_p can be reduced by increasing a, and $H_1 + H_2$. a is maximum when H_2 is zero (that is the rotor is levitated in the field of only one coil, which must obviously be located underneath it at all times).

Setting $H_1 = H_{c1} \approx 1200$ A cm^{-1} (for niobium), we find

$$a_{max} = 8 \cdot 15 \text{ cm.}$$

Thus the minimum precession rate for a solid niobium rotor on Earth is

$$\omega_{pmin} = 0 \cdot 113 \text{ deg day}^{-1}.$$

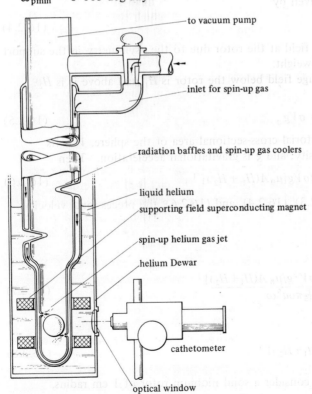

to vacuum pump

inlet for spin-up gas

radiation baffles and spin-up gas coolers

liquid helium

supporting field superconducting magnet

spin-up helium gas jet

helium Dewar

cathetometer

optical window

Figure 16.2.3. A superconducting gyroscope using a solid niobium rotor. Read-out is optional, imperfection in the rotor being viewed to determine the axis of spin of the sphere.

A superconducting gyroscope has been constructed by the Jet
Propulsion Laboratory (JPL) for the purpose of studying the causes of
precession (Harding, 1966).   Figure 16.2.3 shows the general arrangement
of the gyroscope.

The rotor consists of a niobium sphere 1 in in diameter and spherical to
5 μin.   It is annealed at 2000°C in a vacuum of 10^{-8} torr for up to 150 h
to remove gaseous surface impurities.   This treatment increases H_{c1} and
reduces surface losses to such an extent that a duration of spin of up to
two years can be achieved.

The rotor is levitated in a magnetic field of cusped geometry and is set
spinning at up to 200 Hz by jets of helium gas directed on to it
tangentially.   When the rotor is up to speed (which is achieved in 80 min)
the rotor chamber is evacuated to 10^{-5} torr.   This small residual pressure
permits a very small conductive cooling of the rotor to balance the heat
input caused by the incident light needed for read-out.

Read-out in the JPL gyroscope is achieved by sighting a cathetometer
on a pole of the rotor.   The poles show up as the centres of a series of
concentric rings caused by small irregularities on the rotor surface.   The
sensitivity of the read-out system is strongly dependent on the stiffness
and damping of the suspension.   Stiffness can be achieved at the expense
of rotor mass and therefore of inertia.   Damping is effected by linking the
support field with a short circuited turn of copper.   Distortion of the
support field by linear movement of the rotor is damped out by losses in
the copper.

Drift rates in this gyroscope of $0 \cdot 056$ deg h^{-1} have been achieved.

16.3 The superconducting electron linear accelerator†

In Section 5.3 it was shown qualitatively that type I superconductors
remain substantially free of resistance in the presence of alternating fields
at frequencies up to 10^{11} Hz.   This ability of superconductors to support
lossless high-frequency surface currents is exploited in a number of devices,
of which by far the most important and extensively developed is the
superconducting electron linear accelerator.

The principle of the electron linear accelerator (linac for short) is shown
in Figure 16.3.1.   It consists of a series of cavities resonating at a frequency
of about 10^9 Hz.   The electric field in each of these cavities varies
periodically with a phase differing by some multiple of $\frac{1}{2}\pi$ from its
neighbours.   A charged particle travelling through this structure will
experience a sequence of accelerative electric fields.   These successive
electric fields will all act in the same sense upon the particle, if its velocity
is such that it takes half a period of the high-frequency field to travel
from one active cavity to the next.   This demands that the particle should
travel at constant velocity or that the frequency should change as the

† Fairbank and Schwettman (1967).

particle is accelerated. In fact, of course, the frequency of excitation of a resonant structure cannot be changed appreciably during the flight of a particle. What the accelerator does is to increase the relativistic mass of the particle at constant velocity. Because of the limited length of a linear accelerator (as compared with a synchotron) the linac is invariably used to accelerate electrons, which can be introduced into the structure at almost the speed of light. An electron with an energy of 1 MeV has a velocity of 94% of that of light.

The equivalent circuit of the resonant cavity structure is shown in Figure 16.3.2. The currents flowing on the surface of the cavity give rise to a power loss through their interaction with a surface resistance R_s. The ratio of the average energy stored in the magnetic and electric fields per cycle to the energy dissipated in the surface resistance per cycle is called the magnification Q of the cavity. In conventional copper cavities operating at 10^9 Hz, magnifications of about 5×10^4 can be obtained. We can estimate the power which would be required by a group of such cavities to accelerate electrons to 1 GeV. On the assumption of a field gradient of 9 MV m^{-1} and a cavity volume of 1 m^3, the peak energy stored by the electric field would be

$$E_p = \tfrac{1}{2}\epsilon_0 E^2 V, \qquad (16.3.1)$$

where ϵ_0 is the capacitivity of free space [$= (1/36\pi) \times 10^{-9}$ F m^{-1}], E is the voltage gradient in volts per metre ($= 3 \times 10^6$), and V is the volume

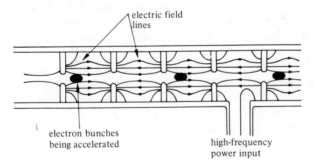

Figure 16.3.1. The basic linear electron accelerator, consisting of a series of resonant cavities through which groups of electrons are accelerated. The time taken for the electrons to travel from one cavity to the next is the half-period of the r.f. source.

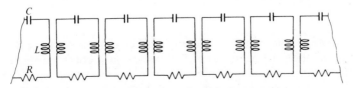

Figure 16.3.2. The equivalent circuit of the linear accelerator.

of the cavity ($= 1/30$ m^3):

$E_p = 12$ J per cavity.

The energy loss per cavity is given by

$$W_L = \frac{E_p f}{Q} = \frac{12 \times 10^9}{5 \times 10^4} = 240 \text{ kW} \ .$$

A 1 GeV accelerator would require 300 such cavities with a total power consumption at 10^9 Hz of 72 MW. Such a prodigious amount of power at high frequency can at the present time only be produced in pulses lasting a few microseconds and repeated every few milliseconds.

Accelerators operated in pulse mode are of less experimental value than those that may be operated continuously, and at the moment only low-energy beams can be produced continuously. However, if the energy loss due to the surface resistance of the cavities could be reduced by a factor of a thousand, continuous operation of a linear accelerator would be possible. This can be achieved by the use of superconductors.

Firstly, however, we should briefly consider one way in which it cannot be done. The d.c. resistivity of pure copper can be decreased by a factor of a thousand by reducing its temperature to within a few degrees of absolute zero. If the high-frequency surface resistance were to decrease similarly, continuous operation of a linac could be achieved. However, because of the anomalous skin effect, the high-frequency surface resistance decreases by a factor of only 7 if copper is cooled to absolute zero. The anomalous skin effect arises in the following way: at very low temperatures the electron mean free path of pure, annealed copper is about 1 mm. A high-frequency magnetic field will, however, only penetrate to a depth much less than 1 mm. This means that an electron may be accelerated by the changing magnetic field at the surface and then drift deep into the interior of the metal out of the influence of the magnetic field. Whilst in this state the electron is effectively lost for surface conduction processes for periods of time which are long compared with the period of a high-frequency field. Thus the apparent high-frequency surface resistance is much greater than the d.c. resistance.

Superconductors do not suffer from an anomalous skin effect because superconducting electrons cannot 'drift' into the depth of a super-conductor. The hypothetical 'normal' electrons can of course suffer an anomalous skin effect but, for the sake of a qualitative argument, we need not consider that in detail.

The theoretical treatment of the surface resistance of a type I super-conductor involves a consideration of the distribution and occupancy of the energy states of the superconductor. The derivation is complex: but we may obtain a qualitative expression by using the two fluid model of Section 5.3 and the electrodynamics of Section 5.2. Thus, as in Section 5.2

we may write

$$J = -\frac{dH}{dx}, \qquad \mu_0 \frac{dH}{dt} = -\frac{dE}{dx}. \qquad (16.3.2)$$

Furthermore, if we assume that the frequency is such that the normal component of current density is very small compared with the superconducting component [see Equation (5.4.7)], then

$$\frac{dJ}{dt} = \frac{n_s e^2}{m} E. \qquad (16.3.3)$$

We are now dealing with sinusoidally varying parameters, so that

$$J(x) = \bar{J}(x) \exp(j\omega t),$$
$$H(x) = \bar{H}(x) \exp[j(\omega t + \alpha)],$$
$$E(x) = \bar{E}(x) \exp[j(\omega t + \beta)], \qquad (16.3.4)$$

where the bar denotes peak amplitudes, and α and β are phase angles.

Substituting these expressions for J, H, and E, we obtain

$$\mu_0 \bar{H}(x) = \lambda^2 \frac{d^2 \bar{H}(x)}{dx^2}. \qquad (16.3.5)$$

This equation is identical in form with Equation (5.2.4). There is no term in j, α, or β indicating that there is no phase shift in H, nor in J, nor in E, at any depth within the superconductor. By contrast the corresponding equation for the penetration of a high-frequency field into a normal conductor is

$$j\frac{\mu_0 \omega}{\rho} \bar{H} = \frac{d^2 \bar{H}}{dx^2}. \qquad (16.3.6)$$

In this the term in j indicates that a phase shift, dependent on X, occurs within the conductor.

In the case of the type I superconductor then, we can write

$$\bar{J} = \bar{J}_0 \exp\left(-\frac{x}{\lambda}\right), \qquad (16.3.7)$$

where $\bar{J}_0$ is the peak amplitude of the surface current density. Now in Section 5.4 on the two-fluid model we derived an expression for the ratio of 'normal' to 'superconducting' current densities. It was

$$\frac{J_n}{J_s} = \frac{\mu_0 \lambda_0^2 \omega}{\rho_n} \frac{t^4}{1 - t^4}, \qquad (16.3.8)$$

where ρ_n is the normal resistivity, λ_0 is the penetration depth at $0°K$, and $t = T/T_c$.

At any particular depth x within the superconductor, the power dissipation is given by

$$W(x) = E(x) J_n(x). \qquad (16.3.9)$$

But

$$E(x) = \frac{m}{n_s e^2} \frac{dJ(x)}{dt} = \frac{j\omega m}{n_s e^2} \bar{J}(x) \ . \tag{16.3.10}$$

Combining Equations (16.3.8), (16.3.9), and (16.3.10), we find

$$\overline{W}(x) = \bar{J}^2(x) \mu_0^2 \lambda_0^4 j\omega^2 \frac{t^4}{1-t^4} \rho_n^{-1} \ . \tag{16.3.11}$$

Now the total loss rate per unit surface area is obtained by integrating $\overline{W}(x)$ between 0 and ∞. Thus

$$\overline{W}_T = \tfrac{1}{2} \bar{J}_0^2 \mu_0^2 \lambda_0^5 j\omega^2 \frac{t^4}{1-t^4} \rho_n^{-1} \ . \tag{16.3.12}$$

The average total power loss is $\tfrac{1}{2}\overline{W}_T$ and the total peak current per unit surface area of the superconductor is $\bar{J}_0 \lambda_0$. Thus the average loss rate per unit surface area is

$$W_T = \bar{I}_0^2 \left(\frac{\mu_0^2 \lambda_0^3 \omega^2}{4\rho_n} \frac{t^4}{1-t^4} \right) \ . \tag{16.3.13}$$

The term in parentheses on the right-hand side is called the surface resistance of the superconductor. The total loss rate in the surface of the superconductor in the presence of an alternating field strength of r.m.s. amplitude H A m^{-1} is then

$$W_1 = H^2 R_s A \ ,$$

where R_s is the surface resistance and A is the surface area. If we take lead at $1 \cdot 85°$K by way of example,

$$\lambda_0 \approx 0 \cdot 5 \times 10^{-7} \text{ m} \ ,$$
$$\rho_n \approx 20 \times 10^{-11} \ \Omega \text{ m} \ ,$$

the surface resistance is

$$R_s \approx 4 \times 10^{-8} \ \Omega \ .$$

By use of a much more accurate theory based on the quantum properties of the superconductor, the surface resistance under these conditions is calculated to be 10^{-8} Ω.

However, despite this discrepancy our relatively simple classical treatment has revealed that the surface resistance varies as $\omega^2 t^4/(1-t^4)$ at low temperatures. It is thus necessary to use the lowest possible reduced temperature in order to obtain the lowest surface resistance.

With the typical value of surface resistance we have derived above, a representative value of the magnification Q can be calculated. Thus

$$Q = \frac{\tfrac{1}{2}\mu_0 H^2 Vf}{\tfrac{1}{2} R_s H^2 A} = \frac{\mu_0 Vf}{R_s A} \ , \tag{16.3.14}$$

where f is the frequency and V/A is the volume/surface ratio. Taking this latter as $0 \cdot 1$ m, f as 10^9 Hz, and R_s as 4×10^{-8} Ω we find

$$Q = 3 \cdot 1 \times 10^9 .$$

This is clearly a considerable improvement in Q over that of copper at room temperature.

The surface resistance is proportional to frequency squared. Hence Q is inversely proportional to frequency. A low frequency is therefore favoured for the superconducting linac. In contrast, the Q of a normal cavity is proportional to root frequency which favours a high frequency.

Equation (16.3.13) shows that the reduced temperature influences the surface resistance strongly. It is desirable to operate a superconducting linac at the lowest practicable reduced temperature. This clearly favours the choice of lead or niobium. The latter allows the lowest reduced temperatures, and, in theory at least, allows the highest voltage gradients to be used.

Only surface currents are exploited in high-frequency superconducting cavities, so that niobium must be used below its lower critical field which is 1380 A cm^{-1}. The maximum voltage gradient is given simply by

$$E_{max} = H_{c1}\left(\frac{\mu_0}{\epsilon_0}\right)^{\frac{1}{2}} , \tag{16.3.15}$$

where μ_0 and ϵ_0 are the electromagnetic constants of free space.
For niobium

$$E_{max} = 0 \cdot 52 \text{ MV cm}^{-1}$$

and for lead

$$E_{max} = 0 \cdot 24 \text{ MV cm}^{-1}$$

both at absolute zero. In fact the maximum voltage gradients are limited by effects other than the lower critical field. Field emission of electrons causes appreciable dissipation at field strengths somewhat less than H_{c1} in both lead and niobium.

The expression for the high-frequency surface resistance of a superconductor as derived by quantum mechanics can be approximated roughly as

$$R_s \propto \frac{\omega^2}{T} \exp\left(-\frac{\epsilon}{2kT}\right) , \tag{16.3.16}$$

where E is the energy gap of the superconductor at temperature T. Both this and our simple expression indicate that the surface resistivity should rapidly decrease towards zero as the reduced temperature falls towards zero. In fact it was found in early tests on lead and niobium that the high-frequency surface resistance reached a residual minimum value as temperature was reduced. This resistance was initially so high as to prevent the realisation of values of Q better than 10^7.

The causes of this residual resistance are many: they are incompletely understood.

The first and most obvious causes are impurities. The preferred method of constructing a superconductivity cavity is now by the electrodeposition of lead onto copper. (The electroplating of niobium is not yet as effective as that of lead.) The purity of the lead layer must be very high. Impurities, especially magnetic impurities, allow local irreversible penetration of the field with a consequent high loss rate. Because the temperature is only about 2°K, many impurities are below their Curie points and therefore magnetic.

Another cause of residual resistance is the circulating current trapped in the lead when it is cooled through its critical temperature. These currents generate fields which tend to be trapped in small normal regions in the lead (cf. the intermediate-state Section 11.2). The currents arise from thermal gradients in the copper–lead structure. Although small, the thermal e.m.f. generated in the copper–lead couples by temperature gradients can drive appreciable currents through the structure.

Other causes of loss in the cavity are the strains caused by the differential thermal contraction between lead and copper. Also the joints between sections and other discontinuities in the structure give rise to residual surface resistance.

The need for a low operating temperature leads to the use of superfluid helium. (This is liquid helium below 2·17°K. In this state it exhibits, among other special properties, very low viscosity and high thermal conductivity.) Because of the high thermal conductivity of this form of helium and because of the very low coefficients of thermal expansion of copper at low temperatures, a mechanically very stable structure is obtained.

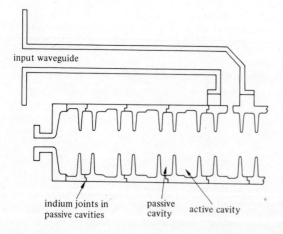

input waveguide

indium joints in   passive   active cavity
passive cavities   cavity

Figure 16.3.3. The nineteen-section superconducting linear accelerator.

Because of its mechanical stability, the structure has a very stable resonant frequency. Coupled with the continuous mode of operation of the superconducting linac, this mechanical stability allows an energy resolution of better than 10^{-4}.

Work on superconducting electron linacs is directed towards the ultimate construction of machines of 1 GeV or higher energy. A model accelerator, 1·5 m in length, has been constructed and is now in operation.

Figure 16.3.3 shows the arrangement of the nineteen resonant sections of the model. Each section consists of one active and one passive cavity. The phase change between one active cavity and the next is π. The sections are joined with an indium seal in the passive cavity. The subdivision of the structure allows the separate lead plating of individual sections: the presence of a passive cavity in each section allows them to be joined with an indium seal without causing a discontinuity in the active surface resistance. The model accelerator operates at 1·85°K and resonates at a frequency of 950 MHz. An energy gradient of 5·0 MeV m^{-1} has been obtained.

References

It will be appreciated that some of these references are of mainly historical interest. They have been included for completeness and so that the reader who would like to trace the development of superconductivity from its remarkable foundations of half a century ago can do so. The total number of references has been restricted and only the most relevant have been cited. The comments appended to various of the references are intended to guide the reader to the most appropriate source of information.

Abrikosov, A. A., 1957, *Soviet Phys.—JETP (Engl. Transl.)*, **5**, 1174.

Anderson, P. A., Kim, Y. B., 1964, *Rev. Mod. Phys.*, **36**, 39.
[This reference gives a good and authoritative coverage of the effects of flux creep and flux flow in type II superconductors.]

Appleton, A. D., Ross, J. S. H., 1969, *Comm. I. Intern. Inst. Refrig., Conf. Low Temp. and Elec. Power, London*, p.129

Bardeen, J., Cooper, L. N., Schrieffer, J. R., 1957, *Phys. Rev.*, **108**, 1175.
[This is the original publication of the microscopic theory of superconductivity. It is a detailed mathematical treatment of the subject. A qualitative review can be found in *Phys. Today*, **16**, number 1, 19–28.]

Bean, C. P., 1962, *Phys. Rev. Letters*, **8**, 250.
[This reference examines the critical field of very small samples in the light of the simple theory given in this book and with reference to the more advanced theory of Pippard which takes account of the non-local effects due to the coherence length.]

Bean, C. P., Swartz, P. S., Hart, H. R., Livingston, J. D., Fleischer, R. L., Buchhold, T. A., Graham, C. D., 1965, *Gen. Elec. Res. Rept.*, number AF-33(657)-11722.

Berlincourt, T. G., Hake, R. R., 1962, *Phys. Rev. Letters*, **9**, 293.

el Bindari, A., 1969, *J. Appl. Phys.*, **40**, 2070.
[In this paper the flow resistivity of a hard type II superconductor is supposed to vary in a more complex way than considered here. Consequently a different expression is obtained for the dimensionless voltage v. However the theory developed here is more accurately applicable to copper clad superconducting wires.]

Bol, M., Fairbank, W. M., 1964, *Proc. 9th Intern. Conf. Low Temp. Phys., Columbus, 1964* (Plenum Press, New York), p.471.

Buchhold, T., 1964, *Cryogenics*, **4**, 212.

Buck, D. A., 1956, *Proc. IRE (Inst. Radio Engrs.)*, **44**, 482.

Burns, L. L., Alphonse, G. A., Leek, G. W., 1961, *IRE (Inst. Radio Engrs.), Trans. Electron. Computers*, **EC-10**, 438.

Chester, P. F., 1967, *Rept. Prog. Phys.*, **30**, 561.

Chirlian, P. M., 1962, *IRE (Inst. Radio Engrs.), Trans. Electron. Computers*, **EC-11**, number 1, 6.

Clark, J., 1966, *Phil. Mag.*, **13**, 115.

Clogston, A. M., 1962, *Phys. Rev. Letters*, **9**, 266.

Cody, G. D., Ed., 1965, *Radio Corp. Am. Techn. Rept.*, number AFML-TR-65-169.
[This report is the best and most complete summary of the compound Nb_3Sn that is available.]

Comm. I. Intern. Inst. Refrig., 1969, *Conf. Low Temp. and Elec. Power, London, 1969*, 16, 47.

Crowe, J. W., 1957, *IBM J. Res. Develop.*, **1**, 295.

Deaver, B. S., Fairbank, W. M., 1961, *Phys. Rev. Letters*, **7**, 43.

Deaver, B. S., Goree, W. S., 1967, *Rev. Sci. Instr.*, **38**, 311.

Doll, R., Nabauer, M., 1961, *Phys. Rev. Letters*, **7**, 43.

Dynatech Corp., 1964, *Rept.*, number APL-TDR-64-71.

Fairbank, W. M., Schwettman, H. A., 1967, *Cryog. Eng. News,* **2,** number 8, 46.
[This reference is a general review of the superconductivity programme at Stanford University. More specific references to the very-high-frequency applications are given in Schwettman *et al.* (1967) and in a report of Stanford University (1967).]
Fietz, W. A., Webb, W. W., 1967, *Phys. Rev.,* **161,** 423.
Giaver, I., 1961, *Discovery,* **22,** 8.
Giaver, I., 1966, *Inst. Elec. Electron. Engrs. Spectrum,* **3,** number 2, 117.
Ginsburg, V. L., Landau, L. D., 1950, *Zh. Eksperim. i Teor. Fiz.,* **20,** 1064.
Goodman, B. B., 1964, *Rev. Mod Phys.,* **36,** 13.
[This volume of the *Reviews of Modern Physics* contains many contributions on the theory of type II superconductors and is excellent supplementary reading.]
Gorter, C. J., 1933, *Archs. Mus. Teyler,* **7,** 378.
Hake, R. R., 1965, *Phys. Rev. Letters,* **15,** 865.
Hancox, R., 1965, *Appl. Phys. Letters,* **7,** 138.
Hancox, R., 1966a, *Proc. 10th Intern. Conf. Low Temp. Phys., Moscow, 1966,* **IIB,** 43.
Hancox, R., 1966b, *Proc. Inst. Elec. Engrs. (London),* **113,** 1221.
Harden, J. L., Arp, V., 1963, *Cryogenics,* **3,** 105.
[This gives a simple summary of the theory of Ginsburg and Landau (1950) and experimental evidence to support its predictions.]
Harding, J. T., 1966, *Jet Propulsion Lab., Tech. Rept.* number 32-897.
Hart, H. R., 1969, *Proc. Summer Study on Superconducting Devices and Accelerators,* BNL50155 (c-55), 1969 (Brookhaven National Laboratory, Upton), 571–600.
Heaton, J. W., Rose-Innes, A. C., 1964, *Cryogenics,* **4,** 85.
[This reference presents evidence that a type II superconductor without defects cannot support a transport current.]
Iwasa, Y., Montgomery, D. B., 1965, *Appl. Phys. Letters,* **7,** 231.
Josephson, B. D., 1964, *Rev. Mod. Phys.,* **36,** 216.
Kantrowitz, A. R., Stekly, Z. J. J., 1965, *Appl. Phys. Letters,* **6,** 56.
Keesom, W. H., 1935, *Physica,* **2,** 35.
Kim, Y. B., Hempstead, C. F., Strnad, A. R., 1962, *Phys. Rev. Letters,* **9,** 306.
Kittel, C., 1962, *Introduction to Solid-State Physics,* second edition (John Wiley, New York), pp.335–371.
Kleiner, W. H., Roth, L. M., Autler, S. H., 1964, *Phys. Rev.,* **133,** 1226.
Kulik, I. O., 1965, *Soviet Phys.—JETP (Engl. Transl.),* **20,** 1450.
Langenberg, D. N., Scalapino, D. J., Taylor, B. N., 1966, *Sci. Am.,* **214,** number 5, 32.
Laverick, C., 1968, *Argonne Natl. Lab. Rept.,* number ANL-315-69.
Little, W. A., 1965, *Sci. Am.,* **212,** 21.
[This article summarises the qualitative reasons for the superconducting energy gap and leads to the prediction of an organic superconducting molecule with high transition temperature. Other texts giving more detailed but concise introductions to the microscopic causes of superconductivity are Kittel (1962) and Weisskopf (1962).]
London, F., 1950, *Superfluids,* volume 1 (John Wiley, New York).
Lynton, E. A., 1964, *Superconductivity* (Methuen, London).
[This book describes in some detail effects, including the exponential variation of specific heat, which are consequential upon the energy gap (see p.11 ff.).
p.48 ff. gives a summary of the theories of the thermodynamic behaviour of type II superconductors.]
Martin, D. H., Bloor, D., 1961, *Cryogenics,* **1,** 159.
Matthias, B. T., 1963, *Rev. Mod. Phys.,* **35,** 1.
Meissner, W., Ochsenfeld, R., 1933, *Naturwissenschaften,* **21,** 787.

Mercereau, J. E., 1967, *Proc. Symp. Phys. Superconducting Devices, Charlottesville, 1967* (Office of Naval Research, Washington) Physics Branch Code 421, paper U1.

Montgomery, D. B., Wizgall, H., 1966, *Phys. Letters*, **22**, 48.

Onnes, H. K., 1911, *Commun. Phys. Lab. Leiden*, number 119b.

Otter, F. A., Solomon, P. R., 1966, *Phys. Rev. Letters*, **16**, 681.

Pontius, R. B., 1937, *Phil. Mag.*, **24**, 787.

Rjabinin, J. N., Schubnikov, L. W., 1935, *Phys. Z. SowjUn.*, **6**, 557.

Rowell, J. M., 1963, *Phys. Rev. Letters*, **11**, 200.

Saint-James, D., de Gennes, P. G., 1963, *Phys. Letters*, **7**, 306.

Schock, K. F., 1961, *Advan. Cryog. Eng.*, **6**, 65.

Schoenberg, D., 1952, *Superconductivity* (Cambridge University Press, London).
[This book contains the thermodynamic theory of superconductivity, the magnetic behaviour of superconductors, and an extensive treatment of the intermediate states. The basic properties are given on p.6 ff.]

Schwettman, H. A., Turneare, J. P., Fairbank, W. M., Smith, T. I., McAshen, M. S., Wilson, P. B., Chambers, E. E., 1967, *Proc. Natl. Particle Accelerator Conf., Washington, D.C., 1967* (Stanford University Report HEPL-503).

Shapoval, E. A., 1962, *Soviet Phys.—JETP (Engl. Transl.)*, **14**, 628.

Silsbee, F. B., 1916, *J. Wash. Acad. Sci.*, **6**, 599.

deSorbo, W., Healy, W. A., 1964, *Cryogenics*, **4**, 257.

Stanford University, 1967, *Rept.*, number HEPL-503.

Stekly, Z. J. J., Woodson, H. H., 1964, *AVCO Res. Rept.*, number 181.

van Suchtelen, J., Volger, J., 1965, *Cryogenics*, **5**, 256.

Swartz, P. S., Bean, C. P., 1968, *J. Appl. Phys.*, **39**, 4991.

Trauble, H., Essmann, U., 1968, *J. Appl. Phys.*, **39**, 4052.

Vetrano, J. B., Boom, R. W., 1965, *J. Appl. Phys.*, **36**, 1179.

Weisskopf, V. F., 1962, *CERN Rept.*, number 62-30.
[See London (1950); the value of the flux quantum, in error by a factor of 2, is given in an inconspicuous footnote on p.152. The error arose because London did not take account of the pairing of superconducting electrons.]

Wilkinson, K. J. R., 1963, *Proc. Inst. Elec. Engrs. (London)*, **110**, 2271.

List of symbols

A magnetic vector potential
a a thickness, a cross-sectional area
B flux density
b a thickness
C specific heat
c a constant
D a diameter
d a diameter
E voltage, voltage gradient
e electronic charge
F a force
f heat dissipation constant, frequency
G free energy
g the gravitational constant, free energy
H field strength
h Planck's constant, a reduced field strength, a height, a heat transfer rate
I current, moment of inertia
i reduced current
J current density
J the Bessel function
j square root of minus one
K thermal conductivity
k Boltzmann's constant, a ratio of currents
L inductance
l a length
M magnetisation, a constant
m electron mass, a constant
N a number density of particles, a constant
n a number of density of particles
P a perimeter, a number of phases
p momentum, reduced thickness, a proportional constant
Q latent heat, a quantity of heat, a magnification
q a surface energy parameter, a cooling parameter, a ratio of currents, a geometrical factor, a quantity of heat
R a creep rate, a radius, a resistance
r a radius
S entropy
s a constant
T temperature, torque
t time, thickness, a reduced temperature
V voltage, volume
v velocity, a reduced voltage
W heat dissipation, a characteristic frequency
w a width, heat dissipation
x distance

α metallurgical phase, a Lorentz force, current gain, a constant
β metallurgical phase, ratio of current densities, a constant
γ coefficient of electronic specific heat, coefficient of heat transfer
δ a distance, the penetration distance
ϵ energy, the energy gap, permittivity of free space
η magnetic viscosity constant
κ the surface energy parameter
λ the London penetration depth, a wavelength
μ permeability of free space
ξ the coherence length
ρ resistivity
σ density, stress
τ a characteristic temperature, a time constant
ϕ magnetic flux, a phase angle, the flux quantum
χ a fraction of superconducting laminae, magnetic susceptibility
ψ an angle
ω angular velocity, a fraction of superconducting electrons,
     the order parameter

Author index